AF263705

The Fleetwood

Engineering Discipline for Software Systems That Must Be Trusted

The Cadillac Fleetwood that AMG Never Built

James L Pulley III

Founded in 2008, Journeyman Publishing's mission is to help professionals maximize their potential. The views expressed in its publications are entirely those of the authors and do not necessarily reflect the views of the staff, owners, or pets of Journeyman Publishing

`https://www.journeymanpublishing.com`

The Fleetwood, Engineering Discipline for Software Systems That Must Be Trusted.

Errata: `Errors@journeymanpublishing.com`

First Edition, 2026.

Casebound:	978-1-964222-02-8
Paperback:	978-1-964222-05-9
Library:	978-1-964222-49-3
e-book:	978-1-964222-06-6

Library editions are manufactured to institutional durability standards.

Journeyman Publishing's logo & the JP Signet are registered trademarks of Journeyman Publishing, LLC. Other product and company names mentioned herein may be trademarks of their respective owners. Rather than use a trademark symbol on everything as it appears, we are using the names in an editorial fashion only to the benefit of the trademark owner. There is no intention to infringe on an existing or retired trademark.

Journeyman Publishing is a specialty publisher for Cybersecurity, IT Professional Services, Software Performance Engineering, Technical Selling, & Software Quality Assurance books. If you have an interest in becoming a published writer, please contact us at newauthor@journeymanpublishing.com

Library of Congress Cataloging-in-Publication Data
Names: Pulley, James L., III, author.
Title: The Fleetwood : Engineering discipline for software systems that must be trusted / James L. Pulley, III.
Description: Pauline, South Carolina : Journeyman Publishing, 2026. | Includes bibliographical references and index.
Summary: Examines authority as an engineering property in software systems, arguing that correctness, reliability, and trustworthiness depend on explicitly declared and enforced governance rather than emergent behavior. Uses the Cadillac Fleetwood and AMG performance engineering as a structural metaphor to analyze time, identity, refusal, legitimacy, and jurisdiction as load-bearing properties of systems. Distinguishes diagnostic reconstruction from authoritative evidence in technical, legal, and executive contexts.
Identifiers: LCCN 2026906831 | ISBN 978-1-964222-02-8 (casebound) | ISBN 978-1-964222-05-9 (paperback) | ISBN 978-1-964222-49-3 (library binding) | ISBN 978-1-964222-06-6 (ebook)
Subjects: LCSH: Software engineering. | Computer systems—Reliability. | Computer software—Quality control. | Computer systems—Management. | Authority (Philosophy) | Risk management—Information technology.
Classification: LCC QA76.9 .P85 2026 | DDC 005.1—dc23

111 Foster Mill Circle, Pauline South Carolina 29374
https://www.journeymanpublishing.com

Trademark, Independence, and Fair Use Statement

Independence and Trademark Notice

The Fleetwood is an independent publication authored by James L. Pulley III. This work is not affiliated with, sponsored by, approved by, licensed by, or endorsed by General Motors Company, Cadillac Motor Car Division, Mercedes-Benz Group AG, Daimler-Benz AG, Mercedes-AMG GmbH, or any of their present or former subsidiaries, divisions, successors, or affiliated entities.

The names Cadillac®, Fleetwood®, General Motors®, Mercedes-Benz®, Daimler-Benz®, AMG®, and Mercedes-AMG®, along with associated vehicle model names, product designations, and brand identifiers, are trademarks or registered trademarks of their respective owners. All such marks remain the exclusive property of their respective trademark holders.

Any references to manufacturers, vehicles, trademarks, engineering practices, historical programs, or product lines are provided solely for purposes of historical reference, technical analysis, commentary, criticism, and identification.

Such references are intended to constitute nominative fair use under applicable trademark and copyright law and are not intended to suggest or imply any sponsorship, authorization, approval, licensing relationship, or endorsement by any trademark holder.

No claim is made to ownership of any trademark, trade dress, or proprietary design referenced in this work.

Engineering and Concept Disclaimer

The vehicle designs, engineering proposals, specifications, performance estimates, and technical interpretations presented in this book represent independent conceptual work and hypothetical engineering studies.

They do not represent official vehicle programs, factory-authorized modifications, approved engineering practices, or product plans of General Motors Company, Mercedes-Benz Group AG, Mercedes-AMG GmbH, Daimler-Benz AG, Cadillac Motor Car Division, or any related entity.

Any resemblance between concepts presented in this work and actual or proposed vehicles is coincidental or arises solely from reference to publicly known engineering practices.

Nothing in this publication should be interpreted as manufacturer-approved guidance for vehicle modification, restoration, or engineering practice.

No Endorsement Statement

Inclusion of brand names, logos, model designations, historical references, or engineering descriptions does not imply endorsement of this publication by any manufacturer or trademark holder.

No manufacturer has reviewed, approved, or authorized the content of this book.

Author's Note

This book is not about a car, and it is not about software. It is about authority as an engineering property.

The Cadillac Fleetwood is used here as a physical reference frame—a machine whose design makes visible the difference between tolerance and enforcement, comfort and truth, accommodation and refusal. AMG is invoked not as a brand, but as an engineering discipline that treated structure, limits, and invariance as prerequisites for power.

Modern software systems struggle with the same problem these machines solved mechanically: preserving truth under load. When time drifts, identity blurs, infrastructure flexes, or jurisdiction becomes implicit, systems may remain operational while losing their ability to testify. They appear stable while becoming structurally dishonest.

This book argues that authority does not emerge from scale, speed, consensus, or tooling. It must be declared, bounded, and enforced. In cars, this happens through frames, brakes, and geometry. In systems, it happens through time, identity, and jurisdiction.

The thought experiment is deliberate. No Fleetwood was ever built this way. But the engineering decisions required to do so are real, historically grounded, and directly applicable to modern systems that must be trusted.

The goal is not nostalgia. The goal is structural clarity.

Audience & Refusal

This book is written for engineers, executives, and system owners responsible for systems that must testify under dispute—financial, operational, or legal. It does not attempt to optimize velocity, maximize convenience, or reconcile competing operational philosophies. Where modern systems privilege accommodation, this book asserts refusal. Where consensus substitutes for structure, this book insists on authority.

Preface to the First Edition

This book was written to establish a position, not to reconcile competing ones.

The arguments presented here are intentionally declarative. They draw boundaries, define terms, and insist on distinctions that are often blurred in modern technical and organizational practice. In doing so, they will feel sharp to some readers. That sharpness is not incidental. It reflects the subject matter.

Authority, governance, and system truth are not negotiated properties. They either exist within declared constraints, or they do not. A system that softens those distinctions in the name of comfort, velocity, or consensus does not become more resilient—it becomes less accountable.

For this first edition, I have chosen to preserve that edge.

Some arguments could be further qualified. Some boundaries could be softened. Some claims could be made more diplomatic. Those choices were deliberately deferred. This volume is intended to function as a foundational articulation of authority as an engineering discipline, not as a comprehensive reconciliation with every prevailing practice or philosophy.

Future editions may refine language, respond to critique, or incorporate additional perspectives as the discipline matures. This edition's role is simpler and more demanding: to state the problem clearly, define the terms rigorously, and allow the implications to stand without dilution.

Readers who disagree are invited to do so precisely and structurally. Readers who agree selectively are encouraged to examine the definitions they reject. In either case, the work asks to be engaged as a whole.

This book does not promise comfort. It promises clarity.

Contents

1

The Frame Is Authority

The Illusion of Stability

The Cadillac Fleetwood does not ask for attention. It assumes it. It is long in a way that modern vehicles have forgotten how to be long. Its hood stretches forward not aggressively, but deliberately. Its rear deck follows behind like a formal procession. The car does not appear to be in motion even when it is moving. It appears to be relocating.

The Fleetwood's authority is visual before it is mechanical. It communicates continuity, permanence, and weight. It looks less like a machine and more like an architectural decision that happens to have wheels.

In the early 1990s, the Fleetwood was one of the last vehicles in America that still embodied the idea that mass itself was a form of safety. It was not a safety in the modern, instrumented sense of crash scores and controlled deformation zones.Thickness of doors. Length of body. Width of track. Height of seating. The message was simple: whatever happens outside, this is larger than it.

This calmness is what most people interpret as stability. And stability, in human perception, is almost indistinguishable from authority.

A system can be perfectly calm and structurally dishonest at the same time.

This is the first illusion.

The Fleetwood feels authoritative because nothing seems to disturb it. It isolates its occupants from noise, vibration, and texture. The driver is not invited to participate. They are insulated. That insulation reads as power.

Inside, the dashboard is wide, horizontal, and calm. The seating is upright and commanding. Visibility is expansive. You do not sit in a Fleetwood. You preside from it. The world passes by in a softened, filtered form. The machine interposes itself between you and reality, and it does so gracefully.

This is not an accident of design. It is the core design objective.

The Fleetwood was engineered to make the environment less intrusive. It does not seek truth from the road. It seeks comfort from it. It reduces variability, not by controlling it, but by averaging it away. It is a system optimized for tolerance.

Tolerance is a powerful illusion.

A tolerant system absorbs error. An authoritative system prevents it.

The Fleetwood's suspension is soft. That softness is not a flaw in its intended mission. It is the mission. The bushings are compliant. The springs are long. The dampers are gentle. Body roll is permitted because body roll is not dangerous at the speeds and expectations for which the car was designed. What matters is that nothing feels abrupt. Nothing feels urgent. Nothing feels out of control.

But what feels controlled is not the same as what is controlled.

This distinction is where the book begins.

Authority is the ability to preserve truth when conditions change, not merely to preserve comfort when conditions remain gentle.

The Fleetwood preserves comfort, not precision. Its chassis yields under load, rolling, transferring, diving—not dangerously, but deliberately. This is refinement, not failure: the system manages deformation rather than resisting it. Authority, by contrast, requires a boundary beyond which the system will not bend, no matter how politely it is asked.

Tolerance and authority are distinct engineering strategies.
This distinction is established here and not re-proven.

The Fleetwood is tolerant in every visible dimension. Its body mounts isolate vibration. Its frame allows subtle twist. Its suspension permits travel. Its steering smooths correction.

These are virtues in its domain. But they are also declarations: this system is designed to accommodate reality, not to enforce structure upon it.

And because it accommodates reality so quietly, it appears stable.

This is where most modern systems make their first mistake. They equate calmness with correctness. They equate lack of visible error with truth. They equate smooth operation with authority.

The Fleetwood teaches us otherwise.

A system can be perfectly calm and structurally dishonest at the same time.

Nothing in the Fleetwood feels unstable. Yet almost nothing in it is rigid. The stability is experiential, not structural. It is achieved through compliance, not through enforcement. It is stability by permission.

When the Fleetwood encounters force, it does not oppose it sharply. It distributes it. It dissipates it. It stretches around it. That is not authority. That is accommodation.

In engineering terms, accommodation is a design strategy. Authority is a design requirement.

Accommodation seeks to survive disturbance. Authority seeks to define boundaries.

This is not a moral judgment against the Fleetwood. It is a clarification of what it is. The Fleetwood is a masterpiece of tolerance engineering. It was built to make large, unpredictable environments feel gentle and navigable. It was not built to define truth. It was built to preserve composure.

Composure is appearance.

Authority is what remains when composure is no longer possible.

The Fleetwood, under sufficient stress, does not assert structure. It yields into softness. It rolls, flexes, and absorbs. That yielding is graceful. It is comfortable. It is safe in the ordinary sense. But it is not authoritative.

An authoritative system does not yield first. It refuses first.

This is the difference that AMG would immediately see.

Where Cadillac engineered tolerance, AMG engineered refusal. Where Cadillac optimized for calmness, AMG optimized for integrity. Where Cadillac allowed deformation as a feature, AMG treated deformation as failure.

The Fleetwood, in stock form, is a visual metaphor for every modern system that looks stable because it is permissive. It does not expose errors because it absorbs them. It does not show conflict because it softens it. It does not reveal truth because it is not designed to preserve it.

The 1990s Performance World

To understand why the Fleetwood and AMG belong in the same sentence, you have to return to a moment when performance engineering was still a craft. Not a brand extension. Not a trim package. Not a marketing adjective. In the early 1990s, performance meant something structural. It meant a small group of engineers taking responsibility for a machine and reshaping its truth.

This was the era before configurators, before corporate performance divisions, before every manufacturer discovered that "sport" was a profitable badge. The performance world was not defined by volume. It was defined by authority. By who was trusted to touch the underlying structure of a car and change it without destroying its identity.

Names like AMG, Alpina, Ruf, Saleen, and Callaway were not decorations. They were signatures. Each represented a workshop that did not simply modify vehicles, but re-authored them. These were not tuning houses in the casual sense. They were engineering authorship firms. When one of these names appeared on a car, it meant that someone had accepted responsibility for the integrity of the entire system.

AMG in this period was not yet a subsidiary. It was not a department inside Mercedes-Benz. It was an independent engineering company that happened to specialize in Mercedes platforms. Its authority did not come from corporate alignment. It came from demonstrated competence. From cars that worked, endured, and outperformed their origins without breaking their underlying logic.

This was the same world that allowed Ruf to be recognized as its own manufacturer. That allowed Alpina to sell vehicles under its own VIN. That allowed Callaway to deliver Corvettes that General Motors itself could not or would not build. These companies existed in the space between factory and fantasy. They were not rebels. They were custodians of mechanical truth operating outside corporate compromise.

Performance, in this era, did not mean speed alone. It meant coherence under stress. It meant that a vehicle remained honest when pushed beyond the conditions its original designers had intended.

That honesty required structure, restraint, and refusal. It required that power be subordinated to geometry, and geometry to integrity.

This is the world in which AMG earned its name.

Affalterbach was not a marketing location. It was a workshop. Cars were built in small numbers. Engines were assembled by individuals whose signatures meant something. The output was limited not by demand, but by how many machines could be built correctly. That limitation was not an inconvenience. It was proof of legitimacy.

An AMG car in this era was not a Mercedes with more power. It was a Mercedes that had been re-asserted. Its tolerances tightened. Its weaknesses confronted. Its structure disciplined. AMG did not add excitement. It removed ambiguity.

Now place Cadillac into this environment.

In the early 1990s, Cadillac occupied a different philosophical space. It was not a performance brand. It was a legitimacy brand. Its job was not to excite. Its job was to represent continuity. Authority. Institutional calm. The Fleetwood was not a sports sedan. It was not trying to be one. It was a formal object. A rolling architectural decision.

Cadillac in this period still believed that luxury meant mass. That safety meant size. That authority meant presence. Its engineering priorities were shaped accordingly. Ride isolation mattered more than lateral grip. Silence mattered more than feedback. Visual gravity mattered more than agility.

And in that, Cadillac was honest.

The Fleetwood was not pretending to be precise. It was not pretending to be agile. It was designed to communicate permanence and composure. It was a vehicle for environments where calm was valued over correctness. Where reassurance was valued over truth. Where yielding was preferred to confrontation.

This is why the Fleetwood is such a powerful object in this book. It is not a failed performance car. It is a successful tolerance machine.

Cadillac optimized for experience continuity. AMG optimized for structural truth.

These are not competing goals. They are orthogonal goals. They exist on different axes. And that is exactly what makes the thought experiment legitimate.

The early 1990s were one of the last moments when this kind of cross-domain engineering was still culturally possible.

It was a time when a company like AMG could have been asked, seriously and without irony, to touch something as formally American as a Fleetwood and bring its own philosophy to bear.

Because in that era, performance firms were still seen as engineering authorities, not brand multipliers.

This is the legitimacy of the thought experiment.

AMG did not begin with engines. It began with constraints. With frames. With geometry. With refusal. It treated power as an amplifier of whatever already existed. If the underlying structure was honest, power would make it authoritative. If the structure was permissive, power would make it dangerous.

Cadillac, by contrast, treated power as an accompaniment to presence. The Fleetwood's engine was not there to redefine the car. It was there to preserve effortlessness. Power was not discipline. It was comfort insurance.

So when we imagine AMG approaching a Fleetwood in 1994, we are not imagining a style exercise. We are imagining a collision of two definitions of authority.

Cadillac authority was ceremonial. AMG authority was structural.

Cadillac authority communicated calm. AMG authority enforced truth.

Cadillac authority absorbed force. AMG authority refused deformation.

This is not a judgment. It is a mapping.

The reason the 1990s matter is because this was the last moment when performance firms still had philosophical autonomy. Before AMG became a brand. Before Alpina became a trim level. Before performance was commoditized into catalog options.

In that window, a firm like AMG could have looked at a Fleetwood and said:

> "This is a powerful object that has never been forced to tell the truth."

And Cadillac, in that era, still possessed enough confidence in its identity to have entertained the question.

The thought experiment is therefore not a fantasy. It is a historically grounded engineering possibility that was simply never exercised.

AMG had already demonstrated that it could take large, conservative platforms and turn them into structurally honest machines without destroying their identity.

The Hammer[1] was not a rebellion against Mercedes. It was an argument for what Mercedes could become if it stopped yielding.

The Fleetwood would have been a similar argument.

Not a sports car. Not a European imitation. Not a provocation.

An assertion.

This chapter does not ask what AMG would have styled. It asks what AMG would have corrected. And that question only makes sense in the 1990s, when performance still meant authorship, and engineering still carried jurisdiction.

This section exists to make the reader comfortable with the seriousness of the exercise.

We are not inventing a car. We are reconstructing an engineering decision that could have been made by people who were still trusted to make such decisions.

That trust is what modern systems no longer possess.

And that loss of trust is exactly what this book is about.

Structural Rigidity as the First Requirement for Truth

Multiple sovereign clocks cannot testify to the same event.

Authority surfaces do not move under observation.
This principle was established mechanically in Chapter 1 and is assumed for all subsequent discussions of time, identity, and jurisdiction.

In physical engineering, those are:

- Fixed datums
- Rigid frames
- Known geometries
- Calibrated instruments

---

[1]https://www.caranddriver.com/reviews/a15141447/mercedes-benz-amg-hammer-archived-test-review/

In systems, their equivalents are:

- A time source that is externally authoritative and auditable
- An identity model that cannot be reinterpreted mid-execution
- A trust model that cannot be overridden for convenience
- An authority boundary that cannot be blurred by aggregation

These are not configuration settings. They are structural members. They must be difficult to change. They must resist optimization. They must survive pressure.

Most modern systems optimize for adaptability. Authority requires the opposite. It requires immovability in specific places.

Self-healing without authority is self-editing. It is a system rewriting its own history to preserve appearance.

A rigid system does not heal by changing what happened. It exposes what broke. That exposure is the price of truth.

A flexible system preserves experience. A rigid system preserves evidence.

You cannot have both everywhere. You must choose where truth is allowed to be uncomfortable.

This is why structural rigidity is the first requirement for truth. Without it, every control surface becomes aesthetic. Logging becomes storytelling. Monitoring becomes theater. Metrics become costume.

We have built systems that are emotionally reassuring and legally indefensible. They feel stable. They feel modern. They feel professional. They cannot testify.

Truth is not a property of intention. It is a property of structure.

And structure must refuse to move.

Frame Members in Systems: Infrastructure, Identity, Time

In the Fleetwood, the frame is not an abstraction. You can stand under the car and point to it. You can put your hand on the rails, the crossmembers, the suspension pickup points, and the engine mounts. These are not styling choices. They are declarations. They decide what is allowed to move and what is forbidden to move.

AMG would not begin a project like this by talking about horsepower. They would begin by crawling under the car and identifying which parts of reality were negotiable and which were not. Which beams were permitted to flex. Which surfaces had to remain invariant. Which attachment points defined geometry as truth rather than suggestion.

That is what "frame" means in engineering. It is the location in the system where interpretation ends.

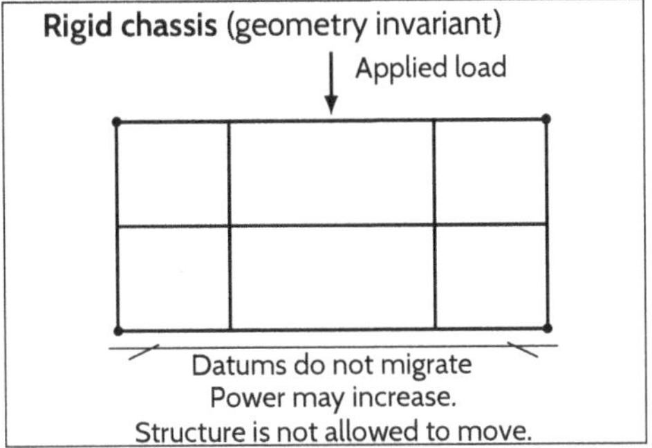

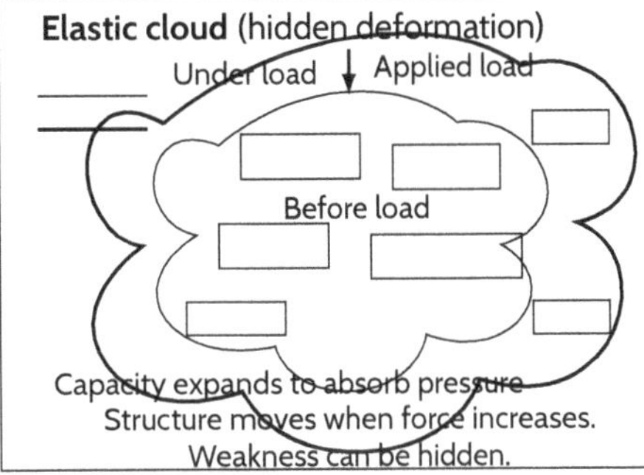

Inversion: a chassis defends geometry under load;
an elastic platform can expand under load and mask deformation.

Figure 1.1: Rigid structures preserve geometry under force. Elastic platforms can "grow" under force—absorbing pressure by moving the frame itself—making weakness quieter rather than corrected.

In a Fleetwood, the frame rail does not grow longer when power increases. The suspension pickup point does not migrate under cornering load. The datum plane does not renegotiate when stress appears.

Power is allowed to increase. Structure is not.

AMG's discipline was always the same: power must reveal structure, never excuse it. If more torque exposed weakness, that weakness was corrected. The frame was not permitted to become more flexible as force increased. That would be a confession of dishonesty.

Now compare that to how modern systems are built.

In cloud infrastructure, the frame is elastic by default. CPU expands. Memory expands. Containers multiply. Queues deepen. Threads replicate. Capacity grows to absorb pressure rather than confront it. The structure is allowed to move precisely when force increases.

This is the inversion of AMG logic.

In the Fleetwood, more power would expose structural weakness. In the cloud, more power hides it.

This is why poorly structured systems grow to fill the available resource pool. Elasticity does not fix dishonesty. It masks it. Bad code scales faster than good truth.

A flexible chassis becomes softer under stress. A flexible system becomes quieter under failure.

Both feel stable. Neither is authoritative.

This is why frame members must be named.

Not as features. Not as services. As beams.

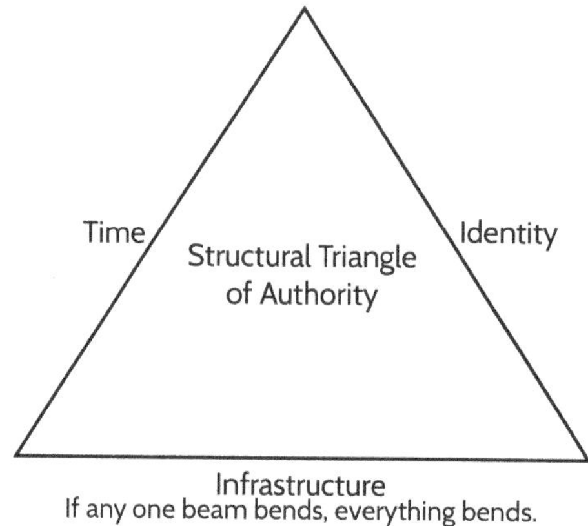

Figure 1.2: Infrastructure, identity, and time are load-bearing members. If any one is treated as flexible, the system becomes interpretive rather than authoritative.

In systems, those beams are:

- Infrastructure
- Identity
- Time

If any one of them bends, everything bends.

Infrastructure, identity, and time are load-bearing authority surfaces. All subsequent chapters assume their non-negotiability and focus only on consequences of violation.

Infrastructure

In a car, infrastructure is the steel. The rails. The crossmembers. The surfaces that define geometry. Virtualization does not exist. You cannot pretend the frame is somewhere else. It is where it is.

In systems, infrastructure is the physical reality underneath abstraction: oscillators, CPUs, memory coherency, power delivery, network latency, packet loss, thermal behavior, failure domains. Virtualization hides these. It does not remove them.

When infrastructure is treated as fluid, the system loses its anchor to physics. Latency becomes negotiable. Failure becomes anecdotal. Performance becomes speculative. The system no longer knows which limits are real.

AMG would never accept a chassis whose rails relocated under load. Cloud systems accept this constantly.

They migrate workloads. They reassign clocks. They reschedule execution. They rebalance networks. And then they ask the system to tell the truth about what happened.

That is a moving frame.

Infrastructure is the beam that carries physical reality into software. If it bends, every metric built on top of it bends with it.

Identity

In the Fleetwood, attachment points define responsibility. When something moves incorrectly, you can trace it to a joint, a weld, a bushing, a mount. Responsibility is mechanical.

In systems, identity plays the same role. Identity answers two questions:

Who is allowed to act? Who is allowed to be believed?

These are not the same. Authentication is not trust. Trust is authority.

If identity is ambiguous, responsibility is ambiguous. If responsibility is ambiguous, truth collapses.

When identity becomes contextual, negotiated, or retroactively interpreted, the system becomes a crowd. Crowds can agree. Crowds cannot testify.

AMG would not accept a suspension mount whose ownership changed mid-corner. Cloud systems accept this every second.

Containers move. Pods restart. Identities are reissued. Permissions mutate. And yet we treat actions taken inside those shifting identities as authoritative.

That is structural negligence.

Identity must be rigid. It must not change because the system is under pressure. It must not be renegotiated for convenience. It must be a load-bearing declaration of responsibility.

If identity bends, accountability bends. If accountability bends, authority dissolves.

Time

In the Fleetwood, there is a datum plane. Geometry exists relative to something that does not move. Without that plane, alignment is meaningless.

Time is that plane in systems. It is not a metric. It is the coordinate system of causality.

But in modern infrastructure, time is elastic.

VM clocks drift. Hypervisors rewrite time. Containers inherit clocks they do not control. Schedulers reorder reality.

A virtual machine does not own its clock. It borrows it. And the lender reserves the right to change it. When the hypervisor enforces correction, history is not synchronized. It is edited.

At that moment:

- Logs become narrative

- Databases write interpretation, not fact

- Ordering becomes negotiable

- Evidence becomes theatrical

This is worse than disagreement. It is revocation of temporal sovereignty.

AMG would never accept a datum surface that moved after measurement. Cloud systems do this by design.

Time becomes elastic precisely where truth demands rigidity.

Together, infrastructure, identity, and time form the structural triangle of authority.

Infrastructure anchors what is physically possible. Identity anchors who is allowed to act and be believed. Time anchors how events are ordered.

These are not features. They are jurisdictional surfaces.

Most modern systems weaken all three simultaneously:

Infrastructure is abstracted and elastic. Identity is contextual and negotiable. Time is borrowed and revocable.

Then we ask the system to testify.

That is like demanding alignment data from a chassis whose rails grow under torque, whose mounts migrate under stress, and whose reference plane floats. The vehicle may move. It may feel smooth. It may even feel powerful. But it cannot tell the truth.

This is why elasticity is so dangerous. It allows poorly structured systems to survive without becoming honest. It lets defects hide inside abundance. It rewards architectural dishonesty by making it comfortable.

In a Fleetwood, increased power would expose weakness. In the cloud, increased power conceals it.

That inversion is the heart of the problem.

AMG insisted that structure must resist force. Modern systems allow structure to retreat from it.

These beams must therefore be declared rigid: Infrastructure must be acknowledged as real. Identity must be anchored and final. Time must be authoritative and auditable.

If any one of them is treated as flexible, the system becomes interpretive. If all three are flexible, the system becomes theatrical.

It can perform stability. It cannot provide evidence.

These are not engineering preferences. They are civil requirements.

Same applied force. Same input. Different testimony.

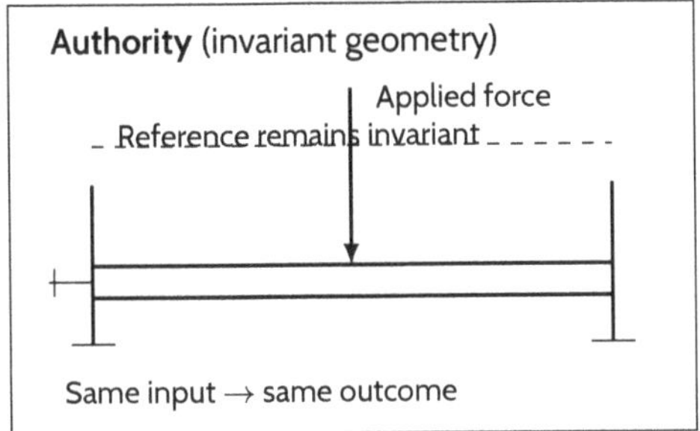

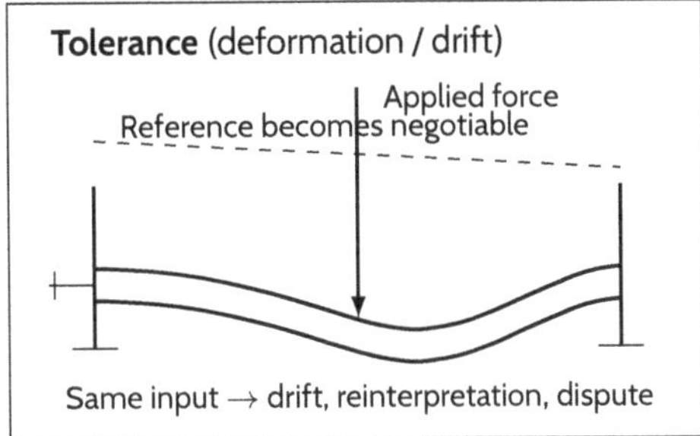

Figure 1.3: Yielding is not safety. Under the same force and the same inputs, an authoritative structure preserves geometry; a tolerant structure deforms and forces interpretation.

What Happens When Authority Flexes Under Load

Governance reference: The structural consequences of authority deformation have already been established. This section examines operational fallout.

In a vehicle, a frame that flexes does not fail immediately. That is what makes it dangerous. The car will still start. It will still move. It will still turn and stop. To an untrained driver, nothing appears wrong. But under load, the behavior becomes unpredictable, and unpredictability is the first signal that authority has been lost.

When a frame flexes, suspension geometry stops being geometry. Camber changes dynamically. Toe shifts under braking. The steering wheel becomes an interpreter rather than a commander. Inputs no longer map cleanly to outcomes. The car feels vague, then nervous, then finally unstable. Not because any single component has failed, but because the reference that defined all relationships is no longer invariant.

Alignment becomes temporary. Handling becomes conditional. Control becomes negotiable.

A flexing frame does not announce itself as broken. It announces itself as inconsistent.

Because once authority begins to deform, everything that depends on it becomes interpretive.

Now look at systems.

When authority flexes in a system, the failure pattern is identical. It does not explode. It becomes vague.

Logs begin to disagree. Metrics begin to contradict each other. Timelines become debatable. Root cause becomes opinion.

The system still runs. The dashboards still light up. The graphs still update. But no two people can describe the same incident in the same way, because the reference frame that would unify their descriptions is no longer rigid.

A flexing authority produces inconsistent logs. Not because logging is broken, but because the clocks, identities, and boundaries that give logs meaning are no longer invariant. Each component reports from its own negotiated reality.

A flexing authority produces false diagnoses. Engineers chase symptoms that are artifacts of measurement rather than causes. Incidents are resolved by suppressing visibility rather than correcting structure. The system becomes quieter, not truer.

This is how organizations develop superstitions about their infrastructure. They stop trusting evidence and start trusting narratives. They rely on seniority, intuition, and pattern recognition because the system itself cannot testify. The system has lost its voice.

A flexing authority produces legal vulnerability. When timelines cannot be reconstructed, when records cannot be ordered, when identity cannot be proven at the moment of action, the system becomes indefensible. Not technically. Civically.

Courts do not accept approximations. Regulators do not accept smoothing. Auditors do not accept consensus. They accept evidence. Evidence requires rigid reference frames. When those frames bend, everything recorded becomes disputable.

This is not hypothetical. It is the reason so many breaches, outages, and fraud cases end in uncertainty rather than resolution. The data exists, but it cannot be trusted. The system cannot speak for itself.

A flexing authority also produces executive deception, often unintentionally.

Dashboards become instruments of reassurance rather than instruments of truth. Metrics are selected because they stabilize narratives, not because they expose reality. Leaders are shown systems that appear calm while structural failures accumulate underneath.

This is not malice. It is structural distortion. When authority bends, reporting follows it. Truth becomes political because structure is no longer mechanical.

The executive is told:

- "Systems are healthy."

- "Performance is stable."

- "Risk is controlled."

What is really being said is:

- "The structure is soft enough that we can hide discomfort."

That is not governance. That is theater.

In a car, a flexing frame eventually produces catastrophe because the loads exceed the structure's ability to adapt. In systems, catastrophe is often postponed indefinitely by scale. More resources hide more dishonesty. More redundancy delays confrontation. More abstraction softens consequence.

But the debt accumulates.

The system becomes larger, more complex, more expensive, and less capable of telling the truth about itself. Each attempt to stabilize it increases the distance between appearance and reality.

That is the cost of allowing authority to flex.

AMG would never accept this. They would never tune a car whose frame was allowed to negotiate. They would never optimize a system whose geometry was conditional. They would insist that before power, before speed, before comfort, the structure must refuse to move.

Because control is not the ability to react. Control is the ability to remain invariant.

A rigid frame produces predictable handling. A rigid authority produces predictable truth.

When the frame does not flex, geometry remains geometry. When authority does not flex, evidence remains evidence.

Once authority is allowed to deform, everything else becomes performance. Logs perform. Metrics perform. Dashboards perform. Meetings perform. But nothing testifies.

This is why the Fleetwood, in its stock form, was stable but not authoritative. It yielded under load. That yielding made it comfortable. It also made it incapable of enforcing structure.

AMG would remove that yielding first. They would not increase power until authority was restored.

Your systems require the same discipline.

When authority flexes, reality becomes optional. When authority is rigid, reality becomes binding.

Control does not come from reaction. It comes from refusal.

The Chassis as an Authority Surface

If Chapter One has argued anything, it is this: authority is not an idea. It is a surface. It is the place where a system refuses to move.

For AMG, that surface would have been the chassis.

They would not have started with horsepower. They would not have started with exhaust, induction, or tuning. They would have started underneath the Fleetwood, on their backs, looking at rails, mounts, seams, and pickup points, asking a single question:

"Where does this car tell the truth, and where is it allowed to improvise?"

In the stock Fleetwood, much of the structure is designed to yield. The frame allows subtle twist. The body is isolated through compliant mounts. The shell is treated as an enclosure, not a contributor to stiffness. Geometry is preserved well enough for composure, but not enforced with the rigidity required for authority. This is not a flaw. It is a design choice. Cadillac built a machine that absorbs force so the passenger does not have to feel it.

AMG would honor that goal, but they would change the hierarchy. Comfort would no longer be the highest authority. Geometry would be.

They would treat the chassis as the primary authority surface: the physical reference frame that every other system must obey. Before power, before speed, before refinement, they would decide what is no longer allowed to move.

The work would be disciplined, not theatrical.

First, they would establish datums. They would define fixed measurement points and measure how the structure behaves under load. Torsional stiffness, bending response, pickup-point migration. Not to create marketing numbers, but to expose where the structure is quietly telling different stories depending on force.

A system that reports different realities under identical inputs is not broken. It is ambiguous. AMG does not build ambiguous machines.

Then they would clarify load paths. Strategic boxing of frame rails. Additional crossmembers. Reinforcement of the places where steering and suspension forces enter the structure. Not everywhere. Only where geometry must be defended. Authority does not mean making everything rigid. It means choosing what must never negotiate.

Next would come the body.

AMG would not turn the Fleetwood into a cage car. That would violate its contract. But they would recognize that the body shell is not neutral. It either lies quietly or tells the truth loudly. Selective stitched welding, reinforcement of key seams, and localized structural participation would be used to make the shell a controlled contributor to stiffness rather than a passive passenger.

Not full seam welding. Not indiscriminate hardening. Only where the system needs the body to testify rather than absorb.

Then the body mounts.

Cadillac uses compliance to preserve comfort. AMG would use firmness to preserve geometry. The mounts would become part of the authority surface. Stiffer bushings, better hardware, tighter tolerances. Still refined. Still civilized. But no longer free to reinterpret the frame under load.

If the body is allowed to drift relative to the frame, geometry becomes theater. AMG would not permit theater.

Suspension pickup points would be reinforced and verified. Steering box mounting would be stabilized. Rear axle location would be disciplined. These are the mechanical equivalents of identity, trust, and time in a system. They define who is allowed to speak, and whose report is believed.

Finally, the whole structure would be validated under load. Not for comfort. For invariance. The question would be simple:

> "When force increases, does truth remain stable?"

Only after the answer was yes would power be allowed to enter the conversation.

This is the mechanical definition of an authority surface. It is the place where deformation is forbidden so that meaning can exist.

Now pull the thread straight into systems.

What AMG is doing to the Fleetwood is exactly what authority engineering must do to modern infrastructure.

They are declaring jurisdiction.

In the car:

- The frame is not allowed to twist.
- Pickup points are not allowed to migrate.
- The datum plane is not allowed to float.
- The steering system is not allowed to lie.

In systems:

- Infrastructure is not allowed to pretend it is infinite.
- Identity is not allowed to be ambiguous.
- Time is not allowed to be borrowed, rewritten, or quietly corrected.

When AMG boxes a frame rail, they are doing what a systems engineer does when they bind execution to known hardware classes, fixed latency domains, and auditable clocks.

When AMG stiffens body mounts, they are doing what a systems engineer does when they remove implicit trust in orchestration layers and make authority explicit at service boundaries.

When AMG stitches welds selectively, they are doing what a systems engineer does when they reinforce specific trust junctions instead of trying to "secure everything" with abstraction.

When AMG refuses to let suspension geometry float, they are doing what a systems engineer does when they refuse to let identity context change mid-transaction.

The mapping is exact because the problem is the same. Authority is the refusal to negotiate reference frames.

Cloud systems, by contrast, are built on the opposite philosophy.

Where AMG says: "Do not let structure move under load," cloud platforms say: "Let structure expand under load."

Where AMG says: "Power must reveal weakness," cloud platforms say: "Power should conceal weakness."

Where AMG says: "Geometry must remain true," cloud platforms say: "Geometry is an implementation detail."

AMG would never accept a chassis that became more flexible as power increased. Modern platforms are designed to do exactly that.

AMG would never accept a datum surface that moved after measurement. Modern virtualized systems rewrite time as a matter of routine.

AMG would never accept pickup points whose ownership changed mid-corner. Modern systems allow identity and execution context to migrate continuously.

AMG would never accept a steering system that lied quietly. Modern dashboards do exactly that when their reference frames are soft.

This is why elasticity is so dangerous. It allows poorly structured systems to survive without becoming honest. It lets defects hide inside abundance. It rewards ambiguity by making it comfortable.

AMG did not build comfortable ambiguity. They built quiet refusal.

Their discipline was simple: Power must expose structure. Structure must refuse deformation. Only then does control exist.

That is why the chassis becomes the first authority surface. It is the first place where truth can be enforced mechanically.

And that is why this book begins here.

Because in cars and in systems, control is not the ability to react faster. Control is the ability to remain invariant.

Control does not come from flexibility. It comes from refusal.

Control begins where deformation ends.

Where Elasticity Is Legitimate & Where It Becomes Dishonest

Elasticity is not the enemy of authority. Elasticity without declared invariants is.

Above the authority surface—where demand is uncertain, experimentation is reversible, and failure carries no external consequence—elastic systems are not only appropriate but efficient. They reduce friction, enable discovery, and allow organizations to defer commitment while information is incomplete.

The failure occurs when elasticity is permitted to cross the authority boundary without constraint. When elastic systems absorb financial, legal, or operational liability without declared invariants, elasticity becomes a narrative device rather than an engineering property. At that point, variability is no longer managed—it is merely displaced.

This book does not reject elasticity. It rejects the pretense that elasticity can substitute for accountability. Authority surfaces require commitments that can be testified to, audited, and defended. Elasticity may operate above those surfaces, but it cannot be allowed to erode them.

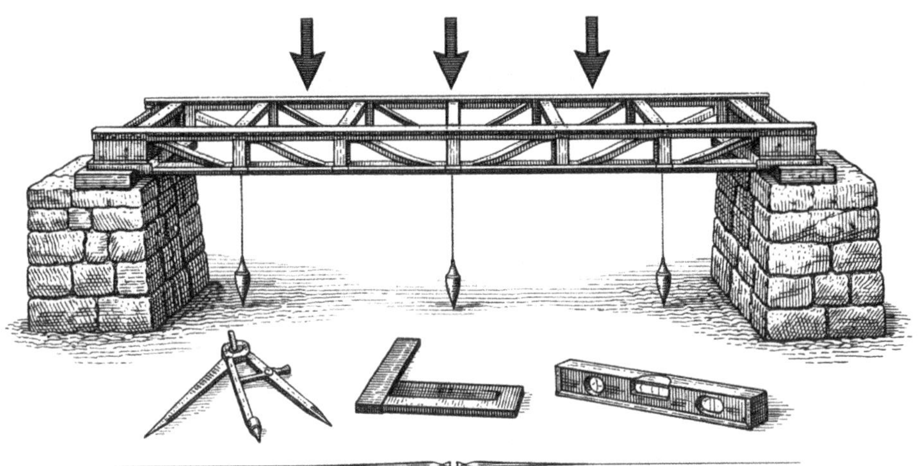

Authority exists where geometry does not negotiate under force.

2

Power Without Brakes Is Delusion: Refusal is the foundation of control.

Braking as an Authority System

AMG would never have evaluated brakes as parts; they evaluated braking as architecture. Braking authority emerges from an integrated control stack that introduces, amplifies, governs, and sustains force predictably—if any layer lies, the system lies. Visible hardware mattered less than making refusal repeatable and independent of chance, with the booster serving as authority amplification so stopping reflected system intent, not human compensation. Hydraulic circuit design and ABS channelization were jurisdiction models: partitioned authority, segmentation, and blast-radius control that allowed refusal without total collapse.

Brake fluid was not maintenance trivia. It was the integrity of the medium through which authority traveled. Moisture reduces boiling point. Heat introduces compressibility. Compressibility introduces ambiguity. Ambiguity destroys refusal. If the medium lies, the system lies. AMG would treat fluid specification, temperature tolerance, and service interval as structural concerns, not consumables. Authority depends on the honesty of its transport.

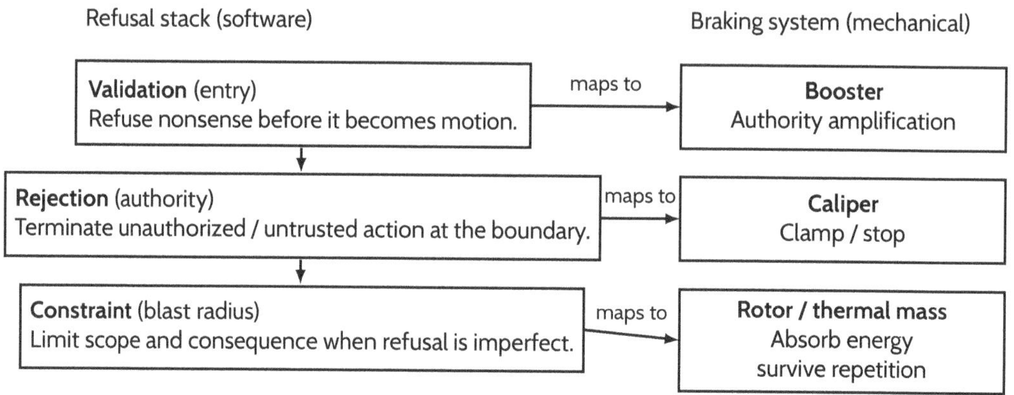

Figure 2.1: Braking is a refusal stack. Validation, rejection, and constraint are structural authority surfaces—analogous to booster, caliper, and rotor mass—governing how motion is refused under stress.

Thermal capacity and recovery defined whether refusal was real or theatrical. One perfect stop proves nothing. Refusal must survive repetition. Otherwise it is coincidence masquerading as control.

So AMG would not ask, "How strong are the brakes?" They would ask, "Is the entire braking stack capable of refusing truthfully under stress, repeatedly, without negotiation?"

That is the question of authority.

Now translate directly.

In systems, we commit the same error that novice builders commit with cars: we look at output hardware and mistake it for control. We count compute cores. We measure throughput. We celebrate acceleration. And we ignore the braking system.

We rarely examine:

- how decisions are amplified,

- how authority is partitioned,

- how trust is transported,

- how time is carried,

- how refusal behaves under repetition.

Validation is the booster. Rejection paths are the calipers. Constraint is the rotor mass. Segmentation is the hydraulic circuit model. Time integrity is the brake fluid. Durability is thermal capacity.

And yet we talk about none of these with the seriousness we reserve for speed.

Just as AMG would never declare a braking system complete by measuring rotor diameter alone, no serious platform can declare itself authoritative by measuring throughput alone. Authority lives in the refusal stack:

How force is amplified. How authority is scoped. How truth is carried. How error is contained. How limits behave when tested repeatedly.

Brakes are not the opposite of performance. They are what make performance honest.

In the physical world, AMG proved this with steel, hydraulics, and heat. In the digital world, the same discipline must be enforced with validation, identity, time, and jurisdiction.

A system that produces endlessly but refuses weakly is not powerful. It is permissive.

A system that refuses cleanly, repeatedly, and without drama has authority.

That is why braking is not limitation. It is proof.

And that is why:

Speed that cannot be refused is not power. It is risk.

Braking as a System of Refusal

What makes brakes authoritative is not their ability to create friction. It is their ability to refuse momentum predictably, repeatedly, and without negotiation.

That is a structural property, not a behavioral one.

If braking were merely an act, then any sufficiently strong intervention would qualify. A wall is very good at stopping a car. So is a tree. So is a ravine. But none of those are brakes. They do not refuse. They merely absorb. A brake is different. A brake is a system that decides when motion ends, how it ends, and under whose authority.

This is why brakes cannot be understood as hardware alone. They are an integrated architecture built around three requirements: heat management, pressure regulation, and repeatability.

Heat management is not a secondary concern. It is the first constraint on honesty. Braking is the conversion of kinetic energy into thermal energy. If the system cannot absorb and dissipate that heat in a controlled way, it loses its ability to refuse. Fade is not inconvenience. Fade is authority collapse. It means the system can no longer enforce its own decisions. The car still has brakes, but it no longer has control.

In an authoritative system, braking performance must be invariant with respect to repetition. One perfect stop is irrelevant. Otherwise refusal is coincidence, not structure.

Pressure regulation is equally fundamental. Braking authority requires proportionality. The system must be able to translate intent into force smoothly, predictably, and without discontinuity. If pressure arrives too abruptly, control becomes violent. If it arrives too slowly, control becomes tentative. If it arrives inconsistently, control becomes interpretive.

This is why the booster, the master cylinder, and the hydraulic geometry matter as much as the caliper. They define how refusal is expressed. They determine whether stopping is a decision or a reaction.

Repeatability ties everything together. A braking system that behaves differently each time under identical inputs is not a braking system. It is a suggestion engine. Authority requires invariance. The same conditions must produce the same outcomes, or else truth cannot exist.

So brakes are not hardware. They are thermodynamics. They are fluid dynamics. They are control theory. They are behavioral invariance.

They are a system of refusal.

This is why AMG never treated braking as an accessory. They treated it as jurisdiction. They asked whether the vehicle was capable of refusing motion under power, under heat, under repetition, and under uncertainty. If it could not, no amount of engine output was meaningful.

Now translate.

In systems, braking is not a component. It is an architectural layer. It exists wherever a system decides what will not be allowed to proceed.

Validation is braking. It is the point at which nonsense is refused before it becomes motion.

Rejection is braking. It is the point at which unauthorized or untrusted actions are terminated deliberately rather than allowed to fail later.

Constraint is braking. It is the point at which scope, scale, and blast radius are limited by design rather than by accident.

Together, validation, rejection, and constraint form the refusal stack. They are the digital equivalent of calipers, fluid, and thermal mass. They convert possibility into control.

Most systems treat these as inconveniences. They are added late, tested lightly, and measured rarely. Throughput is celebrated. Refusal is apologized for.

But just as a car that accelerates well but cannot stop is not performant, a system that produces output but cannot refuse input is not powerful. It is permissive. It is optimistic. It is hoping that nothing bad happens rather than structuring itself to prevent it.

Refusal is what allows a system to preserve its own geometry under force.

In a car, that geometry is suspension alignment, steering feedback, and stability. In a system, that geometry is identity, time ordering, and causal integrity.

When a braking system fails, the car does not merely go faster than intended. It becomes geometrically untrue. The driver no longer knows where authority lives. Control becomes a negotiation with physics.

When a refusal system fails, software does not merely accept bad input. It becomes epistemically untrue. The system no longer knows which actions are legitimate. History becomes narrative. Logs become persuasion. Metrics become reassurance.

Refusal is the preservation of geometry under force and refusal must work at speed. Brakes are tested when everything is moving quickly and consequences compound rapidly. The ability to say "no" when nothing is happening is trivial. Authority is the ability to say "no" when everything is happening.

In systems, this means:

- refusing invalid requests under peak load,
- rejecting unauthorized actions during incident response,
- enforcing constraints when scale pressures temptation,
- preserving identity boundaries when orchestration is chaotic.

Refusal that works only in calm conditions is not refusal. It is theater.

Refusal also prevents amplification of error.

In a car, if braking is weak, small mistakes become large disasters. A slight misjudgment becomes a crash. Momentum amplifies error because nothing arrests it.

In systems, if validation is weak, small faults propagate. A malformed message becomes corrupted state. A confused identity becomes unauthorized access. A timestamp error becomes a legal ambiguity. Without refusal, errors do not stop. They compound.

This is why authority is not the ability to act. It is the ability to prevent action.

Power creates motion. Authority decides whether motion is permitted.

Most platforms are built to say "yes." They are optimized for throughput, elasticity, and accommodation. They measure success in how much they can accept, process, and emit.

Very few platforms are built to say "no" well. Very few treat refusal as a first-class performance metric. Very few budget engineering time for the quality of rejection paths, the clarity of constraint enforcement, or the dignity of failure modes.

This asymmetry is not accidental. Refusal feels like loss. Validation feels like friction. Constraint feels like weakness. Saying "no" rarely produces a dashboard celebration.

But braking systems are never judged by how little they intervene. They are judged by how reliably they stop disaster.

Authority is not the ability to act. It is the ability to refuse.

A system that can do everything but refuse nothing is not powerful. It is reckless. It has no jurisdiction over itself. It is at the mercy of its own momentum.

Braking systems exist to subordinate motion to intent. Refusal systems exist to subordinate capability to authority.

That is their entire purpose.

And until refusal is treated as architecture rather than behavior, no system can claim control.

Authority is the right to refuse motion predictably, repeatedly, and without negotiation.

Validation, Rejection, and Constraint as System Brakes

If braking is the system of refusal, then validation, rejection, and constraint are its working surfaces. They are the places where authority becomes real. They are where a platform stops being a producer of output and becomes a governor of consequence.

Yet most systems treat them as inconveniences.

Validation is seen as overhead. Rejection is treated as error handling. Constraint is viewed as an obstacle to scale.

This is throughput worship: the belief that a system is defined primarily by how much it can accept and how fast it can act. Everything that slows motion is framed as friction. Everything that refuses motion is framed as failure.

In software, we have forgotten that.

Rejection is the second brake.

Validation answers, "Does this make sense?" Rejection answers, "Does this have the right to act?"

Rejection is the refusal of unauthoritative inputs. It enforces identity, permission, and scope. It determines who is allowed to cause motion and who is not.

Most platforms blur this line. They validate structure but fail to validate authority. They ensure a request is well-formed but not well-founded. They confirm syntax while ignoring jurisdiction.

This is the equivalent of a braking system that checks pad thickness but ignores whether the driver has access to the pedal.

Rejection must be explicit, immediate, and decisive. It must terminate motion at the boundary, not after work has already been performed. Delayed rejection is not rejection. It is logging.

A system that accepts unauthorized input and then fails later has not refused. It has narrated its own compromise.

Rejection is the caliper. It clamps authority to reality. It is where trust stops being abstract and becomes mechanical.

Constraint is the third brake.

Constraint exists to limit blast radius. It defines how much damage can occur even when validation and rejection fail. It is the system's acknowledgment that refusal mechanisms are not perfect and that failure must be survivable.

In a car, this is redundancy, partitioned circuits, and localized control. In systems, it is rate limiting, quotas, resource isolation, circuit breakers, sandboxing, and scoped credentials.

Constraint is not pessimism. It is humility engineered into architecture.

A system without constraint assumes perfection. A system with constraint assumes failure and prepares for it. That preparation is authority.

Validation prevents nonsense. Rejection prevents illegitimacy. Constraint prevents catastrophe.

Together, they form the braking system of software.

But notice how rarely these are celebrated.

We celebrate throughput. We celebrate scale. We celebrate latency reduction. We celebrate availability.

We do not celebrate:

- how many bad requests were refused cleanly,
- how many unauthorized actions were stopped at the boundary,
- how small the blast radius remained during failure,
- how consistently the system preserved its own geometry.

Nobody brags about brakes until they fail.

A manufacturer does not advertise, "Our brakes stopped the car today." They advertise horsepower, acceleration, and top speed. Brakes are invisible until they are absent.

Software behaves the same way. Validation and rejection are invisible when they work. Constraint is invisible when it holds. The system appears fast, fluid, and powerful. Only when refusal collapses does anyone notice it was there at all.

And because refusal is invisible, it is underfunded.

Teams budget for compute. They budget for storage. They budget for throughput. They budget for observability.

They rarely budget for refusal paths.

They rarely ask:

- How quickly can we say "no"?
- How consistently do we say "no"?
- How much damage is possible when we fail to say "no"?
- How clear is our refusal when it happens?

Throughput worship teaches us that "more" is always better. But brakes teach the opposite lesson: more power without more refusal is not progress. It is escalation of risk.

A platform that celebrates how much it can ingest but never measures how much it can refuse is like a car that celebrates horsepower but never tests braking distance. It is not confident. It is naive.

Validation, rejection, and constraint are not defensive programming techniques. They are structural authority surfaces. They define the boundaries of legitimacy, the limits of consequence, and the shape of truth inside the system.

Without validation, nonsense becomes real. Without rejection, illegitimacy becomes operational. Without constraint, failure becomes catastrophic.

These are not bugs. They are missing brakes.

And just as no engineer would ship a car whose braking system was "good enough for now," no serious platform should ship without its refusal stack being fully engineered, tested, and budgeted.

Yet most do.

They trust observability to catch problems after they happen. They trust resilience to mask failure after it spreads. They trust scale to absorb mistakes after they multiply.

That is not control. That is hope.

Braking systems are not built on hope. They are built on refusal.

And until software systems treat validation, rejection, and constraint with the same seriousness that engineers treat calipers, fluid, and thermal mass, they will remain fast, capable, and fundamentally ungoverned.

They will produce endlessly.

They will refuse weakly.

And when refusal collapses, no amount of throughput will save them.

Why Platforms Optimize Output but Ignore Refusal

No system is dishonest by accident. It becomes dishonest because the environment around it rewards speed and punishes restraint.

Platforms do not ignore refusal because engineers lack discipline. They ignore refusal because every incentive structure surrounding modern software tells them to. The culture, the metrics, and the economics all align around one principle: motion is success. Anything that slows motion is suspect. Anything that stops motion is failure.

So systems are taught to run, not to govern.

Metrics reward production, not discipline.

What gets measured gets funded, and what gets funded gets engineered. We measure throughput. We measure latency. We measure concurrency, utilization, requests per second, transactions per minute. We build dashboards that glow when motion increases. We alarm when motion declines.

We rarely measure:

- how quickly invalid inputs are refused,
- how consistently unauthorized actions are rejected,
- how small the blast radius remains during failure,
- how often nonsense is prevented from becoming work.

In most organizations, a system that produces more output is celebrated. A system that refuses more input is interrogated.

Refusal does not produce a victory graph. It produces a flat line. It produces absence. And absence looks like failure to cultures trained to worship motion.

In automotive terms, this is like evaluating a car only by its horsepower and never by its braking distance. It is a category error so deep it feels normal.

Revenue follows velocity.

Modern platforms monetize activity. More requests mean more compute consumption. More transactions mean more data movement. More engagement means more revenue. The business model itself is coupled to throughput.

Refusal interrupts revenue flow. Validation slows onboarding. Rejection blocks conversion. Constraint limits scale.

So refusal becomes economically uncomfortable. A platform that says "no" appears to be turning away opportunity. A platform that says "yes" appears inclusive, expansive, and profitable.

But authority is not obstruction. Authority is definition.

Brakes are not a rejection of speed. They are what make speed usable.

Yet platforms are built inside markets where speed is virtue and stopping is suspicion. So refusal is softened, delayed, or hidden. It is implemented quietly and measured rarely.

Refusal feels like loss.

Saying "no" produces immediate negative feedback. Users complain. Support tickets increase. Product teams feel constrained. Executives see friction. So refusal becomes politically dangerous. Acceptance becomes the default virtue.

The ability to absorb abuse is mistaken for robustness. But robustness is not infinite acceptance. Robustness is controlled refusal.

A car that absorbs impact is safe. A car that never brakes is suicidal.

But in software we blur this distinction. We celebrate systems that accept unlimited input and call them resilient. In reality, they are permissive. They have abandoned authority.

Constraint feels like failure.

Limits make systems honest, but honesty is uncomfortable. Rate limits feel like weakness. Quotas feel like lack of confidence. Capacity ceilings feel like poor planning. So they are hidden or deferred.

But constraint is not failure. Constraint is truth.

In a car, braking distance is not embarrassing. It defines the machine. In software, constraint should define authority. Instead, it is treated as a blemish.

Cloud makes saying "yes" cheap and saying "no" rare.

Elasticity inverted engineering discipline.

Historically, refusal was natural because resources were finite. Systems had to say "no" because they physically could not say "yes" forever. Cloud removed that friction. Compute expands. Storage grows. Throughput scales. Limits appear optional.

So power became defined as absorption. The ability to accept everything was mistaken for control.

But absorbing everything is not control. It is surrender.

AMG would never accept a chassis that flexed more under power. Cloud celebrates platforms that flex more under load.

AMG insisted that power expose weakness. Cloud allows power to conceal weakness.

Elasticity makes poor structure survivable. It does not make it honest.

And because these systems survive, their dishonesty becomes normalized. Nothing breaks catastrophically. The dashboards stay green. The business grows. So no one asks where refusal lives.

Until something finally breaks in a way that cannot be masked.

This is why dishonesty is systemic, not malicious.

No one is trying to deceive. They are being rewarded for speed. They are being punished for restraint. They are being taught that acceptance is virtue.

So platforms evolve into engines without brakes.

They look powerful. They scale impressively. They remain fundamentally ungoverned.

And then the curveball appears.

Add a CDN and origin traffic collapses. Tune a WAF and forty percent of traffic disappears. Implement real caching and databases grow quiet.

Dashboards panic.

Because the system just stopped doing massive amounts of work. Not because demand vanished. Because illegitimate, repetitive, and unnecessary work was refused.

Nothing broke. Authority arrived.

This is where caching becomes a governance problem instead of a performance problem.

A real cache strategy forces a system to declare where truth is allowed to live.

Some content is legitimately public. It wants to be repeated freely. Static assets, documentation, marketing pages, product listings, category pages. If the origin is serving these repeatedly, it is wasting authority. Public content belongs at the edge.

So the first declaration of refusal is: **The origin will not answer what the world already knows.**

Then there is content that is not public but still repeatable within an identity boundary. Account dashboards, authenticated views, session-based pages that do not change constantly. These can be held by the client temporarily. Not as performance hacks, but as jurisdictional delegation.

So the second declaration is: **Truth may be delegated briefly, but only inside an identity boundary.**

And then there is the sacred category: **Origin-only operations.**

Cart mutations. Payments. Inventory reservations. Credential changes. Security events. Entitlement transitions.

These must always be decided at the core. They must always speak in the present tense. They are not cacheable because they are not answers. They are decisions.

A mature architecture is not one where everything is fast. It is one where the origin is quiet because it is doing only what it alone is allowed to do.

Public cache is refusal of repetition. Private cache is delegation under constraint. Origin-only operations preserve sovereignty.

And when this hierarchy is enforced, the platform changes character. It becomes quieter. More selective. Less dramatic. More honest.

Which is exactly what happens when AMG finishes the brakes on the Fleetwood.

They are not looking for excitement. They are looking for quiet.

They look at rotor temperatures stabilizing instead of spiking. They look at pedal travel becoming shorter and predictable. They look at repeated stops producing the same response. They look at a chassis that no longer compensates for panic.

The car feels different. Not weaker. Disciplined.

It stops relying on mass and tolerance. It starts relying on refusal.

It only accelerates where it can stop cleanly. It only generates force it can govern.

That is authority.

When your CDN collapses origin traffic, when your WAF removes illegitimacy, when your caches absorb repetition, and when your cores become quiet, you are not regressing.

You are witnessing your AMG moment.

The moment when your system stops being tolerant and becomes authoritative.

When Telemetry Replaces Braking, It Becomes Spectacle

Telemetry is observation. Brakes are control. Confusing the two is one of the most dangerous errors modern systems make.

Telemetry watches failure. Brakes prevent it.

Telemetry tells you what is happening. Brakes decide what is allowed to happen.

When a platform treats observability as a substitute for refusal, it does not become safer. It becomes theatrical. It becomes a fast, detailed, beautifully instrumented narrator of its own collapse.

This inversion is subtle because telemetry feels responsible. Dashboards look professional. Graphs feel like command. Alerts sound like vigilance. But watching something happen is not the same as stopping it from happening. Measurement is not authority. Visibility is not governance.

In a car, telemetry is the gauge cluster: speedometer, tachometer, temperatures, warning lights. These describe the system. They do not change it. A driver who stares at gauges while failing to brake is not controlling the car. They are documenting their loss of control.

In software, we routinely behave as if seeing more clearly is the same as governing more effectively. We add dashboards. We add traces. We add logs. We add metrics. We add observability platforms. And we call this "control."

But none of these stop motion. None of these refuse illegitimate action. None of these preserve geometry. They only describe the moment when geometry is already failing.

Telemetry does not prevent overload. It reports it.

Telemetry does not reject invalid input. It records it.

Telemetry does not constrain blast radius. It annotates the explosion.

This is not a criticism of telemetry. Telemetry is essential. But it is subordinate. It is the speedometer, not the brake pedal. It is the witness, not the judge.

The problem begins when telemetry is mistaken for braking.

When teams believe that seeing more is equivalent to governing more, systems drift toward spectacle. Failure becomes more visible, more understandable, more narratable—and no less inevitable.

Dashboards soothe.

They make chaos feel structured. They give the impression that everything is under watch, which is psychologically close to the feeling that everything is under control. But these are different states.

A well-instrumented system that cannot refuse is like a car with perfect gauges and no brakes. It is exquisitely aware of its speed as it drives off the cliff.

This is why modern outages are often described with extraordinary clarity. We know exactly what happened, in exquisite detail. We have timelines, traces, dashboards, and incident reports that read like forensic novels.

And yet the same classes of failures repeat.

Because observation does not change behavior. Authority does.

Constraints govern

Brakes do not watch motion. They shape it. They reduce energy. They terminate action. They impose geometry. They define where movement ends.

In software, constraint plays this role. Rate limits, quotas, circuit breakers, identity boundaries, validation gates, transactional scopes, and resource isolation are the braking system. They are the only mechanisms that stop systems from accelerating into failure.

Telemetry can tell you a limit has been exceeded. Constraint prevents it from becoming catastrophic.

Telemetry can show you a spike. Constraint flattens it.

Telemetry can describe an attack. Constraint blocks it.

Telemetry can narrate corruption. Constraint contains it.

When constraint is weak or absent, telemetry becomes the primary interface to failure. The platform learns to speak beautifully about its own destruction.

That is spectacle.

Systems become fast narrators of their own collapse.

They generate logs at scale. They emit metrics at high resolution. They trace every dependency. They annotate every error.

And none of it prevents the outcome.

This creates a dangerous cultural pattern: engineering effort shifts toward explaining failure instead of preventing it. Teams become excellent storytellers and poor governors.

Postmortems grow sophisticated. Root cause analysis becomes ritual. Dashboards improve every quarter. Constraints remain shallow.

Because building brakes is harder than building gauges.

Brakes must work under load. Brakes must act decisively. Brakes must refuse.

Telemetry only needs to observe.

This asymmetry makes observability seductive. It is safer politically. It is easier to demonstrate. It produces immediate artifacts. It creates the feeling of progress without forcing hard architectural choices.

Installing constraints, by contrast, feels risky. It might block legitimate traffic. It might reduce throughput. It might anger stakeholders. It might reveal that the system was never safe to begin with.

So organizations invest in visibility and defer authority.

The result is a platform that can describe its own failure in real time, at scale, with confidence—and do nothing to stop it.

There is a reason modern fighter aircraft collapsed a cockpit full of gauges into a handful of glass displays. Raw telemetry at scale does not produce control. It produces cognitive overload. The aircraft must first govern itself so the pilot only has to confirm that governance, not replace it.

The plane does not become safer by seeing more. It becomes safer by refusing more.

Stability is enforced by control systems, not by instrumentation. The cockpit exists to supervise authority, not to compensate for its absence.

In software, we inverted that discipline. We let platforms remain ungoverned and then built massive telemetry stacks to compensate. We created monitoring systems so large that they require their own scaling strategies, their own reliability engineering, their own performance budgets.

In some organizations, the observability platform costs more to operate than the production platform it observes.

That is not sophistication. That is confession.

It says:

> "We do not trust the system to refuse failure, so we will pay to watch it fail
> more clearly."

Worse, it normalizes poor brakes. Instead of asking why refusal is weak, teams improve visibility. Instead of designing constraint, they design dashboards. Instead of enforcing authority, they optimize storytelling.

So telemetry becomes spectacle not because it is useless, but because it is asked to perform the job of braking.

This returns us directly to AMG and the Fleetwood.

Cadillac built instrumentation to preserve reassurance. AMG built braking systems to preserve control.

Both use gauges. Only one governs motion.

AMG would never accept a car whose primary safety system was a warning light. They would not say, "At least the driver will know when they are crashing." They would insist the crash not occur.

Modern platforms routinely do the opposite. They build beautiful dashboards that explain why users were compromised, why databases were corrupted, why availability collapsed, and why trust was lost.

They rarely build the refusal systems that would have prevented those events from becoming real.

Chapter 1 established that authority lives in structure. This chapter shows what happens when we refuse to build that structure and instead build theaters to admire its absence.

We light the stage with dashboards. We script the story with traces. We narrate collapse in real time.

And we call it engineering.

Real engineering refuses to make collapse interesting. It makes collapse rare.

AMG did not build better dashboards. They built better brakes.

Until platforms do the same, telemetry will remain what it has quietly become:

Not a tool of control, but a camera pointed at disaster.

Capability Is Not Authority

Capability is the ability to act. Authority is the ability to refuse.

Those are not adjacent ideas. They are opposites that only appear similar when systems are young, lightly loaded, or forgiven by circumstance. Capability describes what a system can do when everything goes well. Authority describes what a system will not allow when everything does not.

A machine that can do many things is impressive. A machine that knows what it must not do is trustworthy.

This is the distinction that defines engineering maturity.

In cars, capability is horsepower, acceleration, top speed, torque. It is the catalog of motion. It is the list of possibilities. Authority lives somewhere else. Authority lives in brakes, in structure, in geometry, in the refusal to let motion exceed control. Authority is the line that power is not allowed to cross.

A car that can go fast is capable. A car that can decide when fast is no longer allowed is authoritative.

Those two properties coexist only when refusal is engineered as carefully as action.

This is why Chapter 2 has been about brakes rather than engines. Power expands possibility. Braking defines jurisdiction. Without brakes, capability becomes risk. With brakes, capability becomes control.

In systems, the same inversion exists.

Capability is throughput. Capability is concurrency. Capability is elasticity. Capability is scale.

These are the numbers that fill dashboards and justify investment. They describe what a system can do when nothing resists it.

Authority is different.

Authority is validation that refuses nonsense. Authority is rejection that refuses illegitimacy. Authority is constraint that refuses catastrophe. Authority is identity that refuses ambiguity. Authority is time that refuses revision.

Authority is what the system is willing to say "no" to.

A system that can do everything is a system that stands for nothing. It has no jurisdiction. It has no boundaries. It has no shape. It is merely a conduit for motion.

And a system that cannot say "no" cannot be trusted. Not because it is malicious, but because it has no means of protecting its own truth. It cannot preserve its geometry under force. It must accept whatever arrives and hope that observation will save it later.

That is not authority. That is exposure.

This is where the Fleetwood and AMG return one final time in this chapter.

If AMG were building a Fleetwood, they would not begin by designing new brakes in isolation. They would begin by asking a different question:

> "Where does authority already exist?"

They would look first to the OEM parts ecosystem. Heavy sedans. Armored vehicles. Performance flagships. Light trucks designed to stop enormous mass repeatedly and predictably. They would search for components that had already been validated under higher loads than the Fleetwood would ever see.

Larger rotors. Multi-piston calipers. Master cylinders designed for greater pressure stability. ABS systems designed to manage complex refusal scenarios.

They would not treat this as borrowing parts. They would treat it as inheriting authority.

And they would not treat GM as a supplier. They would treat GM as a co-author of truth.

The conversation would not be about style. It would be about load. It would not be about branding. It would be about heat. It would not be about performance. It would be about repeatability.

What temperatures were these systems designed to survive? What fade margins were assumed? What failure modes were acceptable? How much variability was tolerated before authority collapsed?

AMG would be searching for engineering proofs of refusal.

They would be looking for components that had already said "no" under harsher conditions than the Fleetwood would ever face. And once found, they would recontextualize that authority onto a lighter, more disciplined platform.

That is not optimization. That is jurisdiction transfer.

Capability creates motion. Authority governs motion.

AMG's brilliance was not that they made cars faster. It was that they made cars more honest. They ensured that power only existed where refusal was already proven. They refused to let motion outrun structure.

This is exactly what serious systems engineering requires.

Before adding capability, a platform must ask: **Where does authority already exist? Which components have proven refusal under load? Which boundaries have survived stress? Which decisions are already irreversible?**

You do not invent authority if it already exists. You inherit it. You integrate it. You defend it.

This is why cryptographic primitives matter. This is why hardware clocks matter. This is why identity providers matter. This is why kernel enforcement matters.

They are not features. They are authority surfaces that have already been tested under pressure. They are the corporate parts bin of trust.

To ignore them and build your own "capability" on top is to reject inheritance in favor of spectacle. It is to build engines without asking where the brakes will come from.

A system that celebrates what it can do but never documents what it will refuse is advertising capability, not authority.

A system that boasts of scale but cannot explain its limits is showing ambition, not control.

A system that measures throughput but never measures refusal is proving that it values motion over truth.

And motion without truth is not power. It is noise.

AMG would never have built a Fleetwood by asking how much power it could tolerate. They would have built it by asking how much refusal it could guarantee. They would have anchored the entire vehicle around that answer.

That is the difference between capability engineering and authority engineering.

Capability engineering asks: "What can this system do?"

Authority engineering asks: "What will this system never allow?"

Only one of those questions produces trust.

Only one of those questions preserves geometry.

Only one of those questions creates a machine that can testify.

So the chapter ends not with a warning, but with a definition.

A system is not defined by the size of its engine. It is defined by the quality of its brakes.

A system is not defined by how much it can produce. It is defined by what it is willing to refuse.

And therefore:

A system is defined more by what it can refuse than by what it can produce.

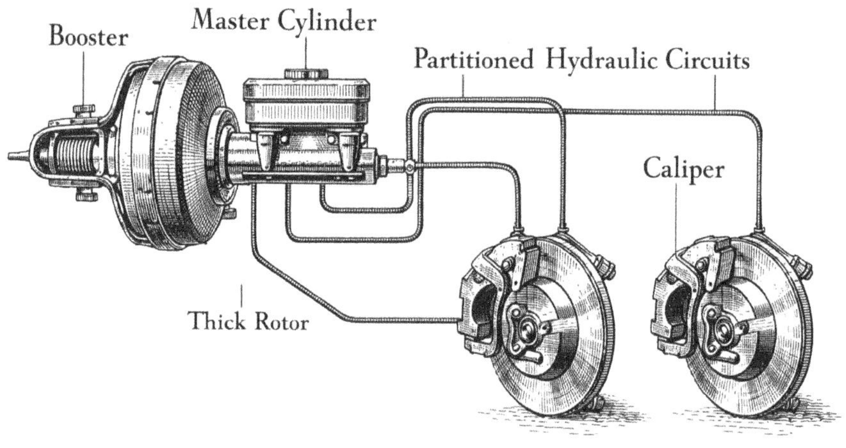

Refusal is an integrated system, not a component.

3

Suspension Is Truth Mediation: Truth requires controlled elasticity.

Suspension as the Translator Between Reality and Structure

The frame is authority. The suspension is interpretation.

The frame defines what is allowed to exist. It establishes geometry, alignment, and permanence. It is the declaration that something has shape and intention. But the frame alone cannot interact with reality. Reality is not static. Roads are not flat. Loads change. Impacts arrive without warning. Without mediation, authority becomes brittle.

That mediation is suspension.

Suspension is not softness. Suspension is not comfort. Suspension is the mechanism that allows authority to remain true while reality moves.

It is the translator between the road and the structure.

A rigid frame bolted directly to the pavement would experience truth in its rawest form: every vibration, every imperfection, every transient shock. That truth would be accurate, but it would also be destructive. The structure would shatter under information density. No interpretation. No survival. Only fracture.

This is brittle truth.

It is correct but unusable. It is precise but terminal. It preserves geometry only until geometry is destroyed.

At the other extreme, a vehicle with no meaningful connection to the road—soft springs, uncontrolled damping, vague geometry—experiences a different kind of truth failure. The structure is protected from reality, but only because reality is being blurred. Forces are absorbed, delayed, and reshaped until they lose causal clarity. The driver feels calm, but calm is not accuracy.

This is dishonest truth.

It survives, but only by misrepresentation. It preserves comfort, but not causality. It protects the structure by lying to it.

Suspension exists to navigate this boundary.

It does not eliminate force. It shapes it. It does not hide reality. It translates it.

That translation must preserve geometry. The structure must still know where it is, what direction it is facing, and how it is loaded. But it must know these things without being destroyed by their immediacy.

This is what makes suspension a truth system, not a comfort system.

AMG understood this distinction instinctively. Cadillac optimized suspension to preserve perception. AMG optimized suspension to preserve geometry.

The Cadillac Fleetwood was tuned to forgive the road. Impacts were softened. Transients were smoothed. Motion was filtered until the experience felt serene. The goal was not to communicate reality, but to insulate against it. The car was meant to float above disturbance.

And it succeeded. The Fleetwood was an achievement of composure. It made chaos feel distant. It made mass feel gentle. It made the world feel cooperative.

But composure is not authority.

AMG approached suspension from the opposite direction. Their goal was not to forgive the road. It was to describe it honestly without allowing it to break the structure. AMG did not ask how little the driver should feel. They asked how much the system must know to remain correct.

Feedback was not discomfort. Feedback was alignment.

The car was allowed to move, but only in ways that preserved geometry. Compliance was engineered, but bounded. Motion existed, but it was disciplined. Every degree of freedom had a purpose, and every purpose had a limit.

This is why suspension is not softness. It is controlled honesty.

A properly designed suspension system is a mechanical admission that truth cannot be delivered raw. It must be buffered, damped, and shaped. But it must not be falsified.

This is the same problem every serious system faces.

The frame establishes authority. The world supplies chaos. Suspension decides how much truth the structure can survive without losing itself.

No suspension means the system shatters under reality. Too much suspension means the system lies about reality.

Trust emerges only when translation preserves geometry.

And in systems, geometry has a name: causality.

As suspension protects geometry in a vehicle, systems engineering must protect causality.

Geometry preserves:

- alignment,
- direction,
- proportional response,
- the relationship between force and structure.

Causality preserves:

- order,
- responsibility,
- sequence,
- the relationship between event and consequence.

If geometry collapses, a car no longer knows where it is pointed. If causality collapses, a system no longer knows what happened.

This is why suspension is the perfect metaphor for truth mediation. It does not protect comfort. It protects correctness under motion. It ensures that translation does not become distortion.

In software, this problem appears everywhere.

Raw reality is overwhelming. Events arrive faster than they can be processed. Signals arrive noisier than they can be trusted. Timing arrives more precisely than it can be preserved.

So systems introduce suspension:

- buffers,
- queues,
- sampling windows,
- rate limiters,
- aggregation layers,
- schedulers.

These are not performance tricks. They are truth mediation systems. They determine how reality is allowed to reach authority.

When they are absent, the system becomes brittle. Everything is synchronous. Every input is immediate. Every failure cascades. The system is technically accurate but operationally fragile. It collapses under the honesty of its own perception.

When they are excessive, the system becomes dishonest. Signals are averaged away. Spikes disappear. Causality is blurred. Events reorder until origin is uncertain. The system feels stable, but only because it no longer knows what is happening.

Rigid suspension breaks truth by fracture. Soft suspension breaks truth by distortion.

Controlled elasticity preserves truth by survivability.

AMG never eliminated compliance. They engineered its limits. Bushings were stiff enough to preserve alignment but flexible enough to prevent shock. Dampers were tuned not for softness, but for response shaping. Geometry was protected even under load.

They did not design suspension to make the car comfortable. They designed it to make the car trustworthy.

That distinction matters.

Comfort is a human preference. Truth is a structural requirement.

Cadillac optimized for perception. AMG optimized for causality.

Both choices are legitimate within their domains, but only one can be called authority.

Because authority must remain correct even when it is uncomfortable.

Suspension, when designed honestly, allows truth to survive motion. It acknowledges that the structure cannot accept reality unfiltered, but refuses to let translation become deception.

This is controlled elasticity.

It is the discipline of allowing movement without allowing distortion. It is the humility of admitting mediation without surrendering causality.

In systems, controlled elasticity means:

- buffering without erasing order,
- damping without hiding spikes,
- sampling without losing sequence,
- aggregation without falsifying origin.

It means that truth is delayed, shaped, and transported—but never rewritten.

A system that transmits raw reality shatters. A system that transmits comfort lies.

Only a system that engineers suspension can be trusted.

Suspension does not protect comfort. It protects geometry.

Bushings, Shock Absorbers, and Geometry as Truth Filters

Every system filters truth. The only question is whether it preserves causality while doing so.

There is no such thing as unmediated reality in a mechanical system. A car does not experience the road directly. The frame never touches asphalt. What the structure knows about the world arrives through bushings, springs, shock absorbers, and geometry. These are not comfort components. They are epistemological components. They decide how truth is allowed to enter the structure.

Bushings are not rubber conveniences. They are buffers.

They decide how much shock is admitted at once. They decide whether force arrives as a spike or as a curve. They decide whether alignment survives impact or whether impact rewrites geometry.

In systems, buffers serve the same role. A buffer determines whether reality arrives as a surge or as a sequence. It protects the structure from instantaneous overload, but it also delays truth. Used correctly, it preserves order while reducing violence. Used poorly, it hides causality.

A bushing that is too soft allows components to drift out of alignment. The wheel still turns. The car still moves. But direction becomes ambiguous. The structure no longer knows exactly where force is coming from or where it is being applied. Comfort is achieved at the expense of accuracy.

A buffer that is too deep does the same thing. Events are absorbed, reordered, or averaged until the system no longer knows what caused what. It feels stable, but it has lost narrative coherence. It can no longer testify.

Bushings do not remove force. They shape its arrival.

Buffers do not remove truth. They shape its visibility.

Shock absorbers (called "dampers" in European engineering literature) are not comfort devices either. They are rate limiters.

A shock absorber decides how quickly motion is allowed to change. It prevents oscillation. It prevents runaway amplification. It prevents resonance from becoming destruction. But it also determines how much of the transient world is allowed to be felt.

In systems, rate limiting plays the same role. It prevents overload. It shapes throughput. It controls acceleration. But it also decides how responsive the system is to real change. Excessive rate limiting flattens reality. Insufficient rate limiting allows chaos.

Shock absorbers do not make a car calm. They make a car stable.

Rate limiting does not make a system slower. It makes a system survivable.

Geometry is the third truth filter, and it is the most sacred.

Geometry preserves relationship. It preserves direction, proportion, and order. It is the difference between motion and meaning.

A car can have perfect bushings and perfect shock absorbers and still lie if geometry collapses. If the suspension arms move in ways that rewrite alignment, the car may feel smooth but it no longer knows where it is pointed. Motion exists, but truth is gone.

In systems, geometry has a name: causality.

Causality is the preservation of:

- event order,
- responsibility,
- dependency,
- before and after.

It is what allows logs to be evidence. It is what allows metrics to be testimony. It is what allows databases to speak.

And causality only exists if time itself is authoritative.

You cannot preserve event order without a clock that refuses negotiation. You cannot assign responsibility without timestamps that mean "before" and "after." You cannot speak of dependency without a stable sequence. You cannot testify without time that is allowed to testify.

Time is not just a measurement. Time is a jurisdiction.

If time drifts, causality drifts. If time is rewritten, responsibility is rewritten. If time is brokered by convenience, truth becomes negotiable.

This is why clocks are not utilities. They are structural members. They are authority surfaces.

A suspension can preserve geometry mechanically, but it cannot preserve geometry if the reference frame itself is unstable. In a vehicle, that reference frame is the chassis. In a system, that reference frame is time.

Without authoritative time:

- logs become narrative,
- metrics become suggestion,
- databases become opinion.

So bushings, shock absorbers, and geometry form a hierarchy of truth filtration:

- Bushings buffer force.
- Shock absorbers shape change.
- Geometry preserves meaning.

- Time anchors causality.

And software has the same hierarchy:

- Buffers absorb load.
- Rate limiters shape throughput.
- Causality preserves truth.
- Clocks preserve causality.

Bad suspension lies by smoothing too much.

When bushings are excessively soft, impact becomes vague. Direction becomes approximate. The structure is protected, but only because information has been destroyed. The system becomes calm at the price of accuracy.

In software this appears as:

- aggressive aggregation,
- heavy averaging,
- long sampling windows,
- alert dampening.

Everything looks stable because the system has stopped perceiving. Spikes disappear. Sequences blur. Responsibility diffuses. The platform becomes comfortable and dishonest.

Rigid suspension lies by breaking under load.

When suspension has no compliance, truth is delivered raw. Every impact becomes damage. Every spike becomes catastrophe. The structure is accurate until it shatters.

In software this appears as:

- synchronous everything,
- no buffering,
- no backpressure,
- no admission control.

The system is honest but fragile. It refuses mediation and therefore refuses survival.

So truth is destroyed in two ways:

By distortion. By fracture.

The first lies. The second dies.

Controlled elasticity preserves truth.

Bushings must be firm enough to preserve alignment and flexible enough to survive impact. Buffers must be shallow enough to preserve order and deep enough to absorb load.

Shock absorbers must allow change while preventing runaway. Rate limiters must admit truth while shaping chaos.

Geometry must be preserved at all costs. Causality must remain inviolate. Time must remain authoritative.

This is where queues appear.

Queues are not performance tools. They are epistemological tools.

They decide:

- what waits,

- what flows,

- what collapses.

A queue that grows without bound is a bushing that has lost stiffness. It no longer preserves geometry. It becomes a swamp of deferred truth.

A queue that is too small is a rigid mount. It fractures under load.

The correct queue preserves causality under stress. It allows truth to arrive without rewriting order.

Sampling windows play the same role.

Sampling is suspension travel. It decides how much reality is observed at once.

Short windows preserve detail but increase volatility. Long windows smooth volatility but destroy resolution.

Neither is universally correct. Both are lies in different ways.

Backpressure is shock absorption.

Backpressure says: "You may not go faster than I can absorb."

It is the mechanical admission that truth must be shaped to survive. Without backpressure, force accumulates until geometry collapses. With excessive backpressure, motion disappears.

Backpressure is not throttling. It is alignment preservation.

So the full translation becomes:

This is not metaphor. It is isomorphism.

Every serious system has a suspension, whether it admits it or not. It has places where truth is delayed, shaped, filtered, and translated. The only choice is whether that translation is honest.

Bad suspension lies by smoothing until causality disappears. Rigid suspension lies by refusing survival.

Trust lives between them.

Suspension Component	System Component	Truth Role
Bushings	Buffers	Shock absorption without distortion
Shock absorbers	Rate limiting	Change control
Geometry	Causality	Meaning preservation
Time reference	Clock authority	Order anchoring
Suspension travel	Sampling windows	Resolution control
Stiffness	Queue depth	Order protection
Shock absorption rate	Backpressure	Stability enforcement

Every platform that claims to value truth must answer four questions:

Where is truth buffered? Where is truth rate-limited? Where is causality preserved? Where is time anchored?

If any of those answers are vague, the system does not have suspension. It has mythology.

Truth does not enter authority raw. Truth must be translated.

But translation must preserve meaning. And meaning, in both machines and systems, is geometry—anchored by time.

Over-Smoothing vs Over-Rigidity in Data Systems

Truth is destroyed in two opposite ways. It disappears when motion is hidden, and it disappears when motion shatters the observer. Between those two failures is where trust lives.

Every suspension system walks this line mechanically. Every data system walks it architecturally. The danger is not imbalance by accident, but imbalance by belief. One side worships calm. The other worships immediacy. Both claim honesty. Both distort truth.

Over-smoothing begins with a reasonable instinct: protect the observer. Reduce noise. Make signals readable. Remove chaos. In moderation, that instinct is correct. Without some smoothing, reality is too violent to survive. But when smoothing becomes the primary design goal, truth is sacrificed to serenity. The system remains operational, but it no longer knows what is happening. It only knows what feels acceptable to display.

Heavy aggregation is usually the first step down this path. Aggregation compresses reality. It takes thousands of discrete events and turns them into a single number. That is not wrong by itself. Compression is unavoidable. But aggregation is destructive. It removes order, removes sequence, and removes responsibility. It answers "how much" while erasing "what happened."

When aggregation dominates, causality fades. You no longer know which event caused which outcome. You only know that something happened somewhere sometime. The platform feels stable because it is no longer capable of perceiving instability.

Rolling averages deepen this effect. A rolling average does not describe reality. It replaces reality with a story about stability. It says that what just happened matters less than what usually happens. Spikes vanish. Transients dissolve. Rare events lose their voice. The system becomes emotionally reassuring and epistemologically useless. It stops speaking in moments and starts speaking in mythology.

Alert dampening finishes the transformation. It begins as a way to reduce noise and prevent operator fatigue. But when applied aggressively, it becomes suppression. Disturbance becomes embarrassing. Silence becomes success. The platform learns that truth is something to be quieted, not understood.

This is how calm dashboards are born. They look mature. They look professional. They feel under control. But calm dashboards often mean the system has stopped perceiving. They are not proof of authority. They are proof of anesthesia.

This is exactly how a luxury suspension behaves. A Cadillac floats because it absorbs disturbance before it reaches the driver. Road imperfections are smoothed. Impacts are softened. Reality is filtered until it becomes comfortable. That is not deception in that domain. It is a legitimate design goal. The car is not meant to describe the road. It is meant to protect the passenger from it.

But if you push that same suspension toward the limit of adhesion, the truth disappears. The structure has no language for urgency. It has no vocabulary for edge conditions. Comfort has replaced geometry. Feedback has been traded for serenity. Authority dissolves quietly.

That is over-smoothing.

Now consider the opposite failure.

Over-rigidity worships raw truth. It rejects mediation. It treats buffering as weakness and latency as dishonesty. It insists that reality must be experienced exactly as it arrives, regardless of consequence. The result is not purity. The result is brittleness.

No buffering means every surge becomes catastrophe. Synchronous everything means every delay becomes deadlock. Fragile pipelines mean every fault becomes collapse. The system is technically accurate and operationally unlivable. It does not survive long enough to speak.

This is the race-car snap.

A race car tuned without compliance communicates reality perfectly, but only until reality exceeds tolerance. When that moment arrives, the system does not adapt. It fails violently. The observer is not misled. They are eliminated. Truth is delivered once, at full force, and then nothing remains to testify.

Over-rigidity claims honesty, but honesty without survival is not truth. It is fragility.

In data systems this appears as pipelines that require perfect input order, architectures that assume infinite reliability, systems that collapse when a dependency stalls, and designs that reject buffering because "latency is evil." These platforms are not wrong. They are simply unwilling to survive.

So the paradox becomes unavoidable. Over-smoothing lies by hiding motion. Over-rigidity lies by making motion fatal. One erases truth quietly. The other destroys truth violently. Both abandon authority.

This is where the AMG Fleetwood comes back into the room.

If AMG were tuning the suspension on the Fleetwood, this trade would be visible immediately. Too soft, and the car would float like its Cadillac ancestor. Impacts would disappear. The road would become emotionally absent. At moderate speed it would feel dignified and composed, but as velocity rose, geometry would become approximate. Direction would blur. Feedback would fade. The car would not lie loudly. It would lie politely.

Too stiff, and the Fleetwood would become brittle. Every expansion joint would feel like a strike. Every transient would threaten structure. The car would be technically accurate and practically unlivable. Truth would arrive raw and destructive. The vehicle would not mislead, but it would not survive long enough to matter.

AMG would accept neither outcome. They would not tune for comfort, and they would not tune for brutality. They would tune for controlled honesty.

The suspension would be allowed to move, but only inside geometry. It would absorb shock without erasing alignment. It would communicate reality without allowing reality to rewrite structure. That is what would make the AMG Fleetwood trustworthy, not impressive.

This is exactly the same choice data systems must make.

Heavy aggregation is luxury float. No buffering is race-car snap.

Rolling averages are insulation. Synchronous pipelines are fracture.

Alert dampening is anesthesia. Zero backpressure is collapse.

One side creates comfortable lies. The other creates catastrophic honesty.

Trust requires a system that can feel motion without being blinded by it and survive motion without denying it. Truth must be allowed to arrive, but it must arrive in a form the structure can endure.

This is why controlled elasticity is not compromise. It is discipline. Buffers must exist, but be shallow. Sampling must exist, but be precise. Aggregation must exist, but be reversible. Alerts must exist, but be honest.

The platform must be able to say, "I will feel this. I will not be broken by this. I will not hide this."

That is authority.

When systems drift toward over-smoothing, they become curators. They select which truths are allowed to be seen. They report tranquility while disorder accumulates beneath. When systems drift toward over-rigidity, they become martyrs. They refuse mediation and collapse with integrity.

Neither can testify.

That is why the AMG Fleetwood is not just a metaphor. It is a working model. Over-smoothing is Cadillac float. Over-rigidity is race-car snap. Authority lives between them. The work of engineering is not choosing calm or brutality. It is designing the suspension that refuses both anesthesia and fracture.

Truth disappears both when motion is hidden and when motion shatters the observer.

Latency, Batching, and Aggregation as Evidence Distortion

Suspension does not only shape force. It shapes time. The moment a system delays truth, it changes it. That is not a flaw. It is a law. Delay is not neutral. Delay is transformation.

In a car, suspension travel is the distance truth must move before it reaches authority. The longer the travel, the more that truth is shaped. The shorter the travel, the more violent the transmission becomes. Neither extreme is honest by itself. One hides motion. The other destroys structure. The work of engineering is choosing how far truth is allowed to travel before it is interpreted.

In systems, latency is that travel.

Latency is suspension distance for information. It is the space between event and awareness. A system with no latency is brittle. A system with too much latency is blind. Both distort truth, just in different directions. Low latency preserves immediacy but risks structural shock. High latency protects structure but risks rewriting causality.

Once truth is delayed, it is no longer raw. It has already been mediated. And mediation always carries consequence.

Batching is the next layer of distortion. In a suspension system, shock absorbers take violent impulses and distribute them over time so the structure can survive. In software, batching does the same thing. It takes many discrete events and turns them into one survivable unit of work. That is not wrong. Without batching, most systems would collapse under peak load. But batching trades precision for stability.

When events are batched, order is weakened. Sequence becomes approximate. Individual responsibility is blurred. You no longer see what happened. You see what accumulated. The system becomes calm because it has stopped experiencing time at human resolution.

Aggregation deepens the transformation. Aggregation removes identity. It deletes individuality. It replaces many truths with one summary. It is not just compression. It is erasure. Aggregation answers "how much" while permanently discarding "which one," "when," and "why."

At that point, truth has become statistical. Useful, perhaps. But no longer evidentiary.

Sampling introduces its own distortion. Sampling is suspension geometry for time. It decides what fraction of reality is allowed to exist at all. Sample too sparsely and you miss causality. Sample too frequently and you overload structure. Either way, reality is shaped by what you choose to notice.

This is where logs lie, metrics mislead, and events reorder. Not because engineers are dishonest, but because the mechanics of mediation are dishonest by nature. Every temporal transformation trades precision for survivability.

The dangerous part is not that this happens. The dangerous part is pretending it does not.

This is where time becomes unavoidable.

In Chapter 1, authority was structure. In Chapter 3, truth is mediated through suspension. But suspension can only preserve geometry if the reference frame itself is stable. In vehicles, that reference frame is the chassis. In systems, that reference frame is time.

Time is not a metric. Time is jurisdiction.

In a physical system, time is singular. On a vehicle CAN bus, every control module shares the same clock domain. Arbitration exists, but sequence is real. "Before" and "after" are not negotiated. They are enforced by physics. That is why automotive diagnostics are possible. That is why fault analysis is authoritative. Logs can testify because time does not argue with itself.

In professional broadcast engineering, the same discipline exists. Cameras, audio mixers, switchers, recorders, and encoders all lock to a shared master clock. Genlock, word clock, black burst, tri-level sync—these are not conveniences. They are sovereignty. A device that drifts is not "handled later." It is rejected. The signal is considered untrustworthy. Broadcast engineers learned early that you cannot reconstruct time after the fact. You must enforce it before the signal exists.

That is authority engineering.

Modern application systems did the opposite. They allowed time to fragment and then built telemetry to compensate.

In a virtualized environment, every virtual machine carries its own clock. That clock is enforced by the local hypervisor, corrected by drift policies, and synchronized only approximately. Even with NTP, time is advisory. Correction is delayed. Authority is hierarchical. What looks like one system is actually a parliament of clocks.

Once time fragments, causality fragments.

Logs no longer agree. Events reorder. Responsibility blurs. Sequence becomes inference.

Modern telemetry tries to rescue this. Distributed tracing, correlation IDs, transaction tags—these are heroic attempts to rebuild causality after time has failed. They create a parallel coordinate system that floats above broken clocks. They do not restore authority. They compensate for its loss.

They exist because timestamps can no longer be trusted.

That means logs are no longer evidence by themselves. They are narrative fragments. They are raw material for reconstruction. They require interpretation before they can testify.

And not every organization has this machinery. Not every system is instrumented for tracing. Not every failure happens inside a cleanly correlated transaction. Which means most systems are already living in a partially fictional timeline.

This is not a software flaw. It is a structural choice.

Mechanical systems enforce time before suspension. Modern systems attempt to reconstruct time after suspension.

That is a downgrade in truthfulness.

Latency changes truth because it moves events away from their moment. Batching changes truth because it merges identities. Aggregation changes truth because it deletes responsibility. Sampling changes truth because it chooses what exists. Clock fragmentation changes truth because it dissolves causality itself.

Every time a system delays truth, it changes it.

That is not optional. It is mechanical.

This brings us back to the AMG Fleetwood.

If AMG were tuning the Fleetwood's suspension, they would care deeply about how far force is allowed to travel before it is interpreted. Too much travel and geometry becomes approximate. Too little travel and the structure fractures. But throughout that tuning, one thing remains fixed: time is singular. Cause and effect are real. Sequence is unambiguous. Truth may be shaped, but it is never reordered.

In modern software, we have inverted that discipline. We shape load with buffers and queues, but we allow time itself to become negotiable. We accept clock drift. We tolerate disagreement. Then we attempt to reconstruct causality after the fact.

That is not suspension engineering. That is archaeology.

The AMG Fleetwood would prove that controlled elasticity only works if authority remains absolute somewhere. Geometry must be preserved. Time must be sovereign. Without that, suspension becomes choreography. Telemetry becomes theater. And evidence becomes storytelling.

Latency, batching, and aggregation are not harmless optimizations. They are acts of translation. And translation, by definition, alters meaning. The question is not whether truth is changed. The question is whether causality survives the change.

In mechanical systems, causality is protected first. In modern application systems, causality is reconstructed later.

That difference defines whether a system can testify—or only narrate.

Designing for Survivable Truth, Not Comfortable Lies

The purpose of suspension is not comfort. Comfort is an outcome that sometimes appears when suspension is doing its job well, but it is never the job itself. The job is survival. More precisely, the job is for truth to survive contact with force, volume, and failure.

Truth that cannot survive load is not truth. It is fragility. Truth that cannot survive scale is not truth. It is a laboratory condition. Truth that cannot survive failure is not truth. It is optimism.

Precision is a dimension of truth, but it is not the whole of it. Precision without survival is useless. A measurement that is perfectly accurate but cannot persist under stress has no authority. It is correct only in moments when it is least needed.

Suspension exists so truth can be both precise and durable. It allows force to be translated without destroying structure. It allows geometry to be preserved even when reality is violent. That is the difference between an instrument and a toy.

This is where Cadillac and AMG diverge philosophically.

Cadillac protected perception. Its suspension was designed to make the road disappear. Impacts were softened. Noise was erased. Disturbance was treated as an insult to comfort. The success of the design was measured by how little the driver felt. Precision was not the objective. Emotional insulation was.

AMG would protect geometry. Their suspension would be designed so the road could be felt without being allowed to rewrite alignment. The driver would not be shielded from reality. They would be informed by it. Truth would arrive shaped, not hidden. Force would be translated, not erased.

Cadillac asked, "How do we make this feel calm?" AMG would ask, "How do we make this remain true?"

That is not an aesthetic difference. It is a moral one.

In a car, comfort is optional. In a system, comfort is dangerous.

Calm dashboards are not proof of stability. They are proof of filtering. They tell you what has been allowed to survive into perception, not what has actually occurred. A dashboard that never disturbs you is not a dashboard. It is a narrative editor.

Stable authority, by contrast, tells hard truth quietly. It does not shout. It does not dramatize. It does not soothe. It simply remains correct under pressure. It continues to speak when load rises. It continues to testify when failure occurs. It continues to order events when clocks drift and dependencies fracture.

That is survivable truth.

In systems, designing for survivable truth means:

- logs must remain causally ordered under load,

- metrics must remain meaningful at scale,

- telemetry must continue to function during partial failure,

- clocks must retain authority even when infrastructure is unstable.

Comfort is irrelevant to all of this. Comfort is not a requirement. Comfort is a temptation.

A system that optimizes for comfort optimizes for silence. Silence is not stability. Silence is absence of testimony.

Precision matters, but only when it is preserved. Precision that collapses under stress is a lie waiting to be told. Precision that cannot survive scale is decoration. Precision that cannot survive failure is vanity.

Survivable truth is precise enough to matter and durable enough to endure.

This is why AMG's approach to the Fleetwood would never be about making it softer or harsher. It would be about making it honest under conditions where honesty normally fails. The suspension would not be tuned to flatter the driver. It would be tuned to preserve geometry when the road tries to destroy it.

That same discipline must exist in system design.

Buffers are not for comfort. They are for continuity. Rate limits are not for politeness. They are for integrity. Sampling is not for convenience. It is for causality preservation. Clocks are not for coordination. They are for authority.

The system must remain capable of telling the truth when:

- traffic spikes,

- dependencies stall,

- infrastructure degrades,

- operators panic.

If it cannot, then it does not have observability. It has storytelling.

Comfort hides truth. Authority preserves it.

Cadillac protected perception because that was its mandate. AMG would protect geometry because geometry is what makes motion meaningful.

Systems must choose the same.

Do they exist to soothe? Or do they exist to testify?

A platform that values comfort will design for silence, smoothing, and emotional stability. It will produce dashboards that feel responsible and professional while truth decays beneath the surface.

A platform that values authority will design for survivability. It will accept discomfort. It will tolerate tension. It will preserve causality even when it is inconvenient. It will allow disturbance to be seen because disturbance is evidence.

Suspension does not protect comfort. Suspension protects geometry.

And in systems: **Observability does not protect feelings. Observability protects truth.**

The real objective is not to make systems feel calm. The real objective is to make truth survive.

Trust Emerges in Controlled Elasticity

Absolute rigidity shatters truth. Absolute softness erases it.

Between those failures is where trust is built.

This is not a philosophical compromise. It is a mechanical one. Every suspension system that has ever worked understood this. Steel alone is too brittle. Rubber alone is too vague. A structure that never yields breaks. A structure that always yields forgets its shape. Truth survives only in a system that moves, but moves within limits that cannot be negotiated away.

Suspension is not softness. Suspension is humility engineered.

It is the admission that reality is stronger than structure, and the refusal to let reality rewrite structure anyway. It is a declaration that force will arrive, that chaos is unavoidable, that disturbance is not a defect. But it is also a declaration that disturbance will not be allowed to define meaning.

Suspension says: "You may act upon me, but you may not become me."

That is the core of trust.

A system that does not move is arrogant. It assumes the world will behave. It assumes correctness is permanent. It assumes purity is sustainable. Those assumptions fail the first time reality refuses to cooperate. When that happens, rigidity does not preserve truth. It destroys the system that could have testified to it.

A system that moves without limits is dishonest. It adapts endlessly. It absorbs everything. It accommodates contradiction. It has no memory of shape. It cannot say when something is wrong, because wrong has been normalized. It does not break. It forgets.

Trust cannot exist in either of these architectures.

Trust requires bounded motion.

Bounded motion means the system yields, but only in ways that preserve identity. It bends, but it remembers its geometry. It absorbs shock, but it refuses deformation. It accepts disturbance without allowing disturbance to rewrite causality.

This is why suspension is the most honest part of a vehicle. The engine boasts. The body persuades. The interior comforts. The suspension tells the truth. It reveals whether the car understands the world or merely insulates itself from it.

AMG never eliminated compliance. They defined where it ended.

That distinction is everything.

They would not turn the Fleetwood into a rigid object that punishes every irregularity. That would be a lie of purity. They would not leave it floating, detached, and emotionally distant from reality. That would be a lie of comfort. They would give it movement that was allowed, measured, and disciplined.

The AMG Fleetwood would move. But it would not drift.

It would yield. But it would not forget alignment.

It would communicate force. But it would not surrender to it.

That is not tuning for performance. That is tuning for truth.

And that is exactly the same discipline systems must learn.

A trustworthy system must move. It must absorb spikes. It must tolerate failures. It must handle volume. It must survive disruption. But it must do all of that without surrendering causality. It must not allow time to become negotiable. It must not allow order to become optional. It must not allow responsibility to dissolve under pressure.

Rigid systems collapse. Soft systems drift. Only bounded systems testify.

This is why elasticity is not weakness. It is humility. It is the recognition that no model is complete and no structure is invulnerable. But humility does not mean surrender. It means knowing where motion is permitted and where it is forbidden.

A suspension system does not debate geometry. It enforces it.

A system architecture must do the same.

Buffers exist so truth can arrive without destroying structure. Limits exist so structure can survive without losing truth.

That is controlled elasticity.

In modern platforms, this is where trust is either engineered or abandoned. If the system absorbs everything, it becomes narrative. If the system absorbs nothing, it becomes brittle. If the system absorbs selectively, it becomes authoritative.

Trust emerges when a system can say:

> "I will move."
> "I will survive."
> "I will not lie."

This is not a feeling. It is a property.

And it is visible.

You can see it in a car that stays aligned after impact. You can see it in logs that remain ordered during failure. You can see it in metrics that still mean something under load. You can see it in timestamps that still testify when infrastructure fractures.

Trust is what remains when motion has occurred and truth is still intact.

That is why suspension is not about comfort. It is about dignity.

It is about refusing both arrogance and surrender. It is about accepting force without abandoning form.

AMG would never remove compliance. They would decide where compliance ends.

That is what authority looks like when it is engineered rather than claimed.

Trust is not rigidity. Trust is truth that survives motion.

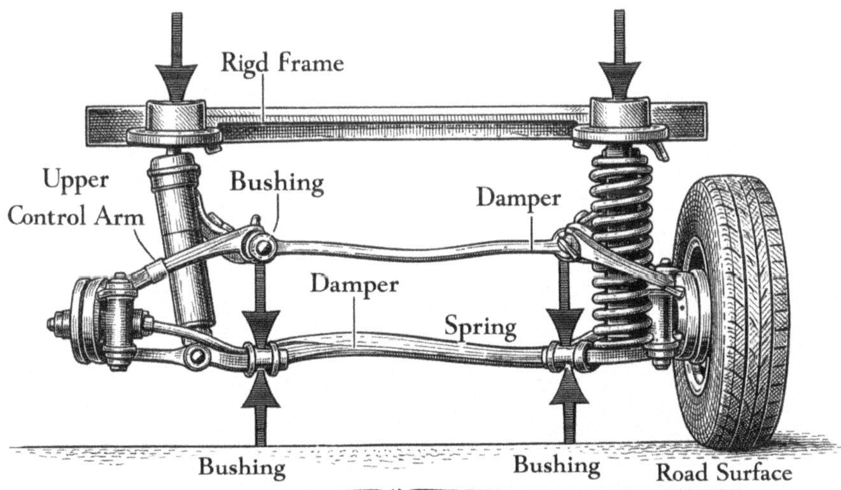

Truth survives only when elasticity is controlled.

4

The Clock is the Speedometer

Instrumentation Before Velocity

Before anyone talks about speed, someone has to answer a simpler question: How fast are we actually going?

That question sounds trivial until you remove the assumption that the instrument answering it can be trusted. If the clock defining a second drifts, and the mechanism recording events within that second also drifts, what does "per second" even mean? If both references are unstable, every rate you compute becomes provisional. You are no longer measuring motion. You are narrating it.

The speedometer is the simplest way to see this.

A speedometer does not create speed. It does not influence direction. It does not absorb force. It is not part of propulsion, control, or stability. And yet, if it lies, every decision made on top of it becomes suspect. Braking distances become estimates. Suspension tuning becomes guesswork. Steering inputs lose grounding. You may still be moving, but you no longer know how you are moving.

Imagine a speedometer driven by a cable whose gear constantly changes. The number of teeth varies. The diameter expands and contracts. Sometimes it slips. Sometimes it binds. It never fails completely. It just drifts.

How fast could you prove you were traveling?

Not how fast it felt. Not how fast you believed. How fast you could demonstrate under scrutiny, to someone who does not trust you.

You could not. And any claim you made about performance would be conditional at best.

This is why instrumentation comes before velocity in serious engineering.

AMG would never increase power on a platform that could not tell the truth about itself. Before adding speed, they would recalibrate the gauges. Before tuning suspension for higher loads, they would ensure the car could report what it was experiencing accurately. Not because instrumentation is glamorous, but because it is jurisdictional. A lying gauge does not merely misinform the driver. It invalidates the engineering conversation.

In the early 1990s, this discipline was visible. Instrumentation was mechanical. Gauges were analog. Tolerances were physical. Calibration was tangible. Drift could be seen, felt, and diagnosed. Because it was visible, it was treated as a defect, not a nuance.

Calibration was authority.

If the needle did not return to zero, something was wrong. If the gauge read inconsistently, it was repaired. If the instrument could not be trusted, the car was not ready.

Modern vehicles did not escape this requirement when the cable disappeared. They re-encountered it in a different form.

In a contemporary car, speed is no longer measured directly. It is inferred. Wheel speed sensors emit pulses. Pulses are counted over time. Speed is calculated as distance divided by time. The gauge no longer observes motion. It computes it.

That computation lives on a shared bus.

On a CAN bus, multiple control units publish and consume data, but they do so inside a single timing domain. Arbitration exists, but time itself is not negotiated. Every module agrees on what a millisecond is. If that agreement breaks down, the failure is visible. The system degrades loudly. Diagnostics become unreliable. Safety systems retreat.

Modern cars solved this by centralizing time before centralizing computation.

Now consider the alternative.

If each control unit applied its own notion of "now," if each corrected drift independently, if time were advisory rather than enforced, the car would still calculate speed. The display would still move. But speed would no longer be provable. The number would be plausible, not defensible.

The car would still drive. Authority would quietly disappear.

This is exactly where modern software systems find themselves.

We speak casually about requests per second, transactions per second, events per second. We graph them. We alarm on them. We present them to executives. *All of these claims assume that the second is stable and that the mechanism recording events against it is authoritative.*

In virtualized environments, that assumption is false more often than we admit. Each virtual machine inherits time from a hypervisor. Each hypervisor enforces correction policies. Drift is tolerated. Synchronization is approximate. Time is no longer enforced; it is negotiated.

The system does not crash. The numbers continue to move. But the coordinate system they depend on becomes elastic.

At that point, the system has not failed. Authority has.

You cannot govern what you cannot measure truthfully. And you cannot measure truthfully if your coordinate system is unstable.

AMG would never trust speed before trusting the speedometer. Modern systems often do the opposite. They optimize throughput while assuming time will behave.

This chapter begins here because everything that follows depends on it.

Time as the Coordinate of Causality

Every system eventually has to answer a simple question: what happened first? Until that question is answered, nothing else can be trusted. Not value. Not volume. Not frequency. Not even correctness.

Time is not a metric layered on top of data. It is the coordinate system that makes data intelligible at all.

Causality requires order. An effect must follow a cause. A dependency must be satisfied before it is consumed. Responsibility must be assignable to a moment, not just an outcome. "Before" and "after" are not narrative conveniences. They are structural requirements. Remove them, and events may still occur, but meaning collapses.

This is why time outranks everything else we measure.

Value without time is just a number. Volume without time is just accumulation. Frequency without time is just repetition.

None of them explain why something happened. They only describe that something happened, stripped of sequence and responsibility.

This is easiest to see in the Fleetwood.

If you are evaluating an AMG-tuned Fleetwood, knowing that it reached 120 mph is less important than knowing *when* it did. Was that speed reached before braking, or after a correction? Was it sustained, or momentary? Did the chassis settle before the maneuver, or did it oscillate through it? The same numeric value can describe discipline or danger depending entirely on its position in time.

Speed without sequence is theater. Speed with sequence is evidence.

AMG engineers would never accept a performance report that did not preserve order. They would not ask only *how fast* the car went. They would ask *when, relative to what,* and *under what conditions*. Because only then does the data become actionable. Only then does it testify.

This is the difference between logs and stories.

Logs are testimony. They assert that something occurred at a specific moment relative to other moments. They can be cross-examined. They can be correlated. They can be challenged. They either hold together under scrutiny or they do not.

Narration is something else. Narration says, "this is what we believe happened." It may be plausible. It may even be accurate. But it is not provable. It relies on reconstruction and interpretation. It is useful, but it is not authoritative.

Time is what separates the two.

A log without trustworthy time is not testimony. It is a diary entry. A metric without ordering is not a witness. It is an impression.

This is why time is the coordinate of causality rather than just another field in a record. When time is stable, events line up. Dependencies become visible. Responsibility can be assigned. When time drifts or fragments, events reorder. Causes appear to follow effects. Independent actions collapse into coincidence. Systems begin to contradict themselves without realizing it.

Nothing looks broken. Everything simply becomes unverifiable.

Metrics suffer from this quietly.

A metric answers questions like "how many," "how often," and "how much." Those are useful questions, but they are subordinate to "when." Without time, a metric can only describe a shape, not a cause. It can tell you that error rates increased, but not what preceded the increase. It can tell you that throughput dropped, but not what triggered the drop. It can tell you that something changed, but not why.

This is why metrics are witnesses, not judges.

A witness without a reliable sense of time is unreliable, no matter how confident it sounds.

In the Fleetwood, this distinction is physical. If a suspension component fails, the order in which forces arrived matters. Did the bump precede the steering input, or did the steering input precede the bump? Did the oscillation build before braking, or after? Without that ordering, diagnosing the failure becomes guesswork. With it, geometry tells a story that can be verified.

AMG's work depends on this. They tune not just for outcomes, but for sequences. They care about how the car transitions, not just where it ends up. That is why they can defend their decisions. They know not only what happened, but when it happened relative to everything else.

Modern systems often lose this discipline.

They generate enormous volumes of data. They measure everything continuously. They report values with impressive precision. But when time is unstable, all of that detail floats free. Events are logged, but not ordered. Metrics are captured, but not sequenced. Systems appear observable while becoming increasingly unverifiable.

At that point, data becomes noise. Not because there is too much of it, but because it no longer agrees with itself.

As this problem became unavoidable, systems engineering responded by building a second layer of order above the first.

Distributed tracing, deep diagnostics, and open telemetry frameworks exist to reassert sequence in environments where local time can no longer be trusted. Correlation IDs, spans, and traces are not merely observability features. They are jurisdictional tools. They attempt to impose logical order on events whose timestamps have already become suspect.

In effect, they say: *If we cannot trust when something happened, we will at least try to trust what happened before what.*

Tracing systems often establish sequence without establishing defensible time. They can tell you that request A called service B, which then called service C, even if the timestamps on those events disagree. They reconstruct causality by relationship rather than by clock.

They still report duration — but that duration is measured on a clock whose authority is already compromised.

The locally logged event carries a timestamp assigned by a local clock. That clock may drift. It may be corrected abruptly. It may be enforced by a hypervisor with its own notion of "now." The event is real, but its reported time is provisional.

The centrally captured trace may report span durations and sub-durations, but those numbers reflect local experience of time, not elapsed time aligned to a single authoritative coordinate. The durations look precise. They are useful. But precision is not authority.

So which truth is correct?

The local log preserves content but may lie about when. The trace preserves order but cannot guarantee sovereign time.

Neither, by itself, establishes full authority.

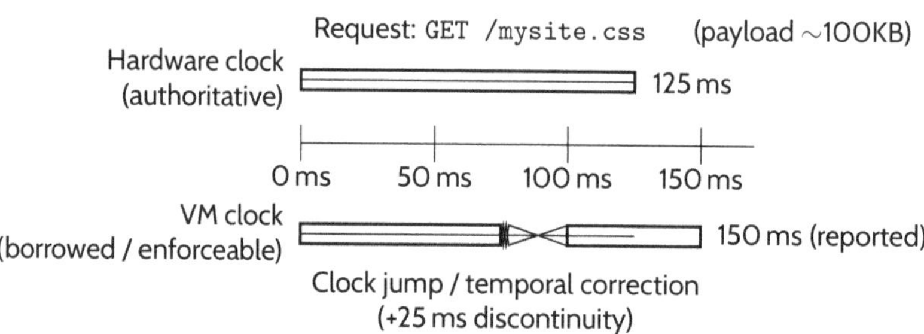

Figure 4.1: Same physical work, different testimony: a VM clock can insert a discontinuity that inflates reported duration without changing elapsed time.

This is not a failure of tooling. It is the consequence of architectural compromise. Once a system allows time to fragment, no downstream instrumentation can fully restore it. Tracing can reconcile order. It cannot retroactively impose a single clock.

This is analogous to forensic reconstruction after a crash. Investigators can determine the order of impacts even if the dashboard clock shattered. They may estimate durations based on distance and deformation. But they cannot prove exact timestamps. They reconstruct causality, not time.

In that sense, deep diagnostics and open telemetry are forensic instruments, not authority surfaces. They are essential, but compensatory. They exist because the system has already accepted that time is no longer singular.

Tracing can reassert order. It can report local duration. It cannot reassert authoritative time.

And that distinction matters.

Because authority requires more than plausibility. It requires defensibility.

AMG would never accept a diagnostic system that could only infer sequence after the fact. Instrumentation would be designed so order was enforced at the moment of observation. Telemetry would confirm, not reconstruct. Logs would testify, not speculate.

Modern systems often reverse that relationship. They allow time to drift locally, then depend on centralized observability to reconstruct order later. This works operationally. It does not work juridically.

Without time, data is noise. With reconstructed order but floating duration, data is plausible. Only with sovereign time does data become evidence.

That is why time is not just another metric. It is the coordinate on which every other truth depends.

Clock Drift as Silent Systemic Failure

Clock drift does not announce itself. It does not trip alarms. It does not halt execution.

It simply erodes agreement.

This is why clock drift is more dangerous than clock failure. A failed clock is loud. A failed clock stops reporting. A failed clock forces intervention. Drift does none of these things. Drift allows the system to continue operating while quietly losing the ability to agree with itself.

Clock failure breaks systems. Clock drift dissolves them.

A failed clock produces gaps. Engineers notice gaps. They investigate. They repair. A drifting clock produces continuity. The system keeps running. The dashboards stay green. The logs keep filling. Nothing looks wrong until someone asks a question the system can no longer answer.

That question is usually some version of: what happened first?

Clock drift begins as a rounding error. A millisecond here. A few microseconds there. No one cares, because nothing obvious breaks. But systems do not reset between events. They accumulate state. Every small disagreement compounds. Over time, independent clocks stop sharing a common present. They inhabit overlapping but incompatible timelines.

The system has not failed. It has fractured.

Once clocks disagree, ordering becomes unstable. Events that occurred later may appear earlier. Dependencies look inverted. Effects precede causes. Retries appear before failures. Compensations run before transactions complete. The logs do not lie maliciously. They lie mechanically.

This is how impossible timelines are created.

A request appears to finish before it starts. An error appears before the operation that triggered it. A recovery appears before the failure it corrected.

Each individual record looks plausible. Together, they contradict reality.

This is why drift is silent. There is no single error to point to. There is no moment where the system "breaks." Instead, explanation becomes harder. Postmortems stretch longer. Engineers argue about interpretation. Teams stop trusting their own data without being able to say why.

At this point, systems often appear successful.

Traffic is flowing. Revenue is posting. Availability is high.

And yet, when something goes wrong, the system cannot explain itself.

Distributed logs disagree. One service insists an event occurred at 10:01:02. Another insists it occurred at 10:01:00. Both are correct locally. Neither is authoritative globally.

When stitched together, the story becomes incoherent.

Metrics contradict reality. Latency graphs show improvement while users complain of slowness. Error rates drop while support tickets rise. Throughput increases while downstream systems choke. The numbers are accurate within their own timelines. The timelines do not align.

Events appear to occur before their causes. Alerts fire before thresholds are crossed. Scale-up actions appear before load arrives. Rollbacks trigger before deployments complete. Each subsystem behaves correctly according to its own clock. The system as a whole behaves irrationally.

This is not a software bug. It is a loss of shared time.

Clock drift corrodes accountability because accountability depends on order. Responsibility requires sequence. You cannot assign fault if you cannot establish precedence. You cannot prove negligence if you cannot prove timing. You cannot defend decisions if you cannot demonstrate what information was available at the moment those decisions were made.

This is where drift becomes dangerous beyond engineering.

From an operational standpoint, teams begin to distrust evidence. Logs are treated as hints. Metrics become suggestions. Traces are consulted for narrative coherence rather than proof. Decisions shift from evidence-based to intuition-driven, not because people prefer intuition, but because the evidence no longer holds together.

From a legal and forensic standpoint, the consequences are worse. If timestamps cannot be trusted, chain of custody weakens. If event order is ambiguous, liability expands. If timelines contradict, testimony becomes vulnerable. The system may have behaved correctly, but it cannot demonstrate that correctness.

Clock drift does not break systems. It breaks accountability.

This is why clock drift is so rarely addressed directly. It does not hurt performance immediately. It does not show up as downtime. It does not trigger customer complaints on its own. It reveals itself only when explanation is required.

And explanation is always required eventually.

AMG would recognize this failure mode immediately in the Fleetwood. If sensors reported values against drifting clocks, diagnostics would become unreliable. If the order of events could not be trusted, tuning decisions would be suspect. The car might still drive well, but engineers would no longer be able to explain why. At that point, refinement stops. Authority is lost.

Modern systems often accept this loss without naming it.

They tolerate drift because it is inconvenient to eliminate. They compensate with correlation. They reconstruct order after the fact. They tell plausible stories. But plausibility is not authority.

Clock failure demands repair. Clock drift invites rationalization.

That is why it persists.

The danger is not that systems behave incorrectly. The danger is that they behave correctly without being able to prove it. When that happens, trust becomes a matter of confidence rather than evidence.

And confidence is not durable.

Clock drift does not crash systems. It lets them keep running while quietly erasing their ability to testify.

That is the most dangerous failure mode of all.

Virtualization and Inherited Temporal Lies

Modern systems rarely keep their own time. They inherit it.

This is not a convenience. It is an architectural decision with consequences that most systems never examine. Virtualization did not just abstract compute, memory, and storage. It quietly abstracted time, and in doing so, it separated execution from temporal authority.

That separation is where truth begins to fray.

In a physical system, time is local and enforced. A crystal oscillates. A counter advances. Drift exists, but it is bounded, visible, and correctable at the source. The system owns its clock, and responsibility for its accuracy is unambiguous.

Virtualization breaks that ownership.

A virtual machine does not own time. Its clock is enforced by the hypervisor. That enforcement may be periodic, corrective, or opportunistic. It may pause time. It may jump time. It may smear corrections slowly or apply them abruptly. From inside the VM, time appears continuous. From outside, it is being negotiated.

The VM executes instructions. The hypervisor decides when those instructions are said to occur.

This is not a bug. It is how virtualization works.

Containers inherit this problem one layer further removed. A container does not even have the illusion of a clock. It inherits time from the host kernel, which inherits it from the hypervisor, which inherits it from a physical clock it does not directly expose. By the time an application reads "now," it is several abstractions away from authority.

Time has become indirect.

NTP does not restore sovereignty here. NTP is advisory. It is a suggestion to move closer to a reference, not an enforcement of truth. Corrections are delayed. Drift is tolerated. Large jumps are avoided because they break applications. The system chooses continuity over correctness.

That choice is understandable. It is also consequential.

Drift correction policies exist specifically to hide instability. They smooth corrections so applications are not startled. They avoid stepping time because stepping time exposes disagreement. Instead, they smear error slowly, allowing clocks to converge without obvious disruption.

The result is a system that feels stable while being subtly wrong.

Time, in these environments, is not enforced. It is negotiated.

Each VM becomes a small republic of time. It has its own laws, its own correction policy, its own sense of "now." These republics coexist on the same hardware while disagreeing quietly about order. They do not argue loudly. They drift politely.

This is how one system becomes many clocks.

From an architectural diagram, it still looks like a single system. Requests flow. Responses return. Metrics aggregate. Logs collect. But underneath, temporal agreement has fractured. Causality is no longer observed. It is inferred.

Inference is not authority.

This is where inherited temporal lies emerge. No component is lying intentionally. Each is reporting time as it understands it. The lie exists at the boundary, in the assumption that these reports are comparable without qualification.

They are not.

An event logged at 10:01:05 in one VM and an event logged at 10:01:04 in another may not have occurred in that order. A span reported as taking 200 milliseconds may not overlap cleanly with a dependency reported as taking 150 milliseconds. The numbers are precise. The ordering is not.

Time looks consistent locally. It is inconsistent globally.

This is why causality becomes inferred rather than enforced. Engineers reconstruct order using correlation IDs, spans, and context propagation. They rebuild sequence from relationships instead of timestamps. This works well enough to debug. It does not work well enough to prove.

The system can tell a coherent story. It cannot guarantee that the story is temporally true.

This is the same failure mode that would make the AMG Fleetwood unacceptable to engineer seriously. If each subsystem in the car measured time independently, if braking events were timestamped differently than steering inputs, if suspension responses were recorded against shifting clocks, diagnostics would become interpretive rather than factual. The car might still drive beautifully. But engineers would no longer be able to explain why.

AMG's authority would be compromised, not because the car failed, but because its explanations could no longer be trusted.

Modern systems accept this compromise routinely.

They delegate time authority without consent. The application does not choose its clock. The container does not choose its clock. The VM does not choose its clock. Authority is passed downward implicitly, along with the assumption that it will be "good enough."

This is the same pattern Chapter 1 warned about.

Authority surfaces can be delegated without consent.

When that happens, the system still functions. What it loses is the ability to assert truth under scrutiny. It becomes operationally effective and evidentiary weak.

Virtualization made scaling easy. It also made time ambiguous.

That ambiguity is survivable most of the time. It is catastrophic precisely when explanation matters most: during incidents, audits, disputes, and failures that require accountability. At that moment, inherited time reveals itself as provisional.

The system runs. The clocks disagree. The truth becomes conditional.

Modern platforms often mistake this for an acceptable tradeoff. They assume that because everything works most of the time, authority is intact. It is not. Authority has been diluted, not destroyed, and dilution is harder to detect than absence.

A failed clock demands repair. An inherited clock invites trust it does not deserve.

This is why virtualization must be understood as a temporal authority problem, not just a resource one. Compute can be overcommitted. Memory can be ballooned. Storage can be thin-provisioned. Those compromises are visible. Time compromise is quieter. It hides behind precision.

And precision is not truth.

When one system no longer means one clock, causality becomes probabilistic. When causality becomes probabilistic, accountability becomes fragile. And when accountability is fragile, authority exists only by assumption.

That assumption holds until someone asks the system to prove itself.

At that point, inherited temporal lies stop being theoretical. They become decisive.

This is not an argument against virtualization. It is an argument for naming its cost honestly. Time is no longer owned. It is borrowed. And borrowed authority is always conditional.

Modern systems inherit time they do not control. They also inherit the consequences of that choice.

Telemetry as Reconstruction, Not Truth

Governance reference: Telemetry observes outcomes; it does not define causality. This distinction is assumed here.

Modern observability systems did not arise because systems became easier to understand. They arose because systems became harder to explain.

When clocks are authoritative and shared, events arrive already ordered. Logs testify. Metrics align. Causality is visible without interpretation. In that world, telemetry is confirmatory. It adds context, not structure.

In modern systems, telemetry has become structural because the structure beneath it has weakened.

Correlation IDs exist because timestamps cannot be trusted to line up. Span IDs exist because local durations cannot be compared cleanly across systems. Transaction tagging exists because event order must be inferred rather than observed. These tools do real work, but the work they do is forensic, not authoritative.

They answer the question, what likely happened? They do not answer the question, what provably happened, when?

This becomes clearer when we separate two kinds of clocks.

Physical clocks attempt to measure elapsed time against a reference. They drift. They are corrected. In virtualized environments, they are enforced by layers that value continuity over correctness. Their authority is conditional.

Logical clocks do something different. They do not measure time at all. They measure order. A logical clock says that event B followed event A, regardless of when either occurred in absolute terms. Logical clocks preserve sequence without claiming duration.

Tracing systems rely heavily on logical clocks.

A trace can tell you that a request entered a system, called three services, waited on two dependencies, and returned a response. It can do this even when the timestamps on those events disagree. It reconstructs causality by relationship rather than by clock.

The trace will still report duration—but that duration is measured on clocks whose authority is already compromised. The numbers are precise. They are useful. They are not sovereign. They describe local experience of time, not elapsed time aligned to a single, defensible temporal reference.

So the trace knows order, and it reports duration, but it cannot guarantee that the reported duration means the same thing everywhere.

This is the uncomfortable truth most observability platforms avoid stating explicitly: tracing trades temporal authority for causal plausibility.

That trade is often the right one. Without it, modern distributed systems would be nearly impossible to operate. But it is still a trade, and pretending otherwise creates false confidence.

This is why observability dashboards often feel persuasive but brittle. They tell a coherent story about how a system behaved. They show flows, bottlenecks, and hotspots. They give teams confidence that they "understand" the system. Most of the time, that confidence is operationally sufficient.

Until explanation becomes defense.

When an incident must be reconstructed under scrutiny—when auditors ask questions, when lawyers examine logs, when executives must testify—narrative coherence is no longer enough. At that moment, the question is not whether the story makes sense, but whether it can be proven.

Tracing systems are optimized for understanding, not for evidence.

They assume consistent context propagation, universal participation, intact instrumentation during failure, and tolerance for approximate time. Those assumptions hold during normal operation. They erode precisely when failure occurs.

This is why telemetry feels strongest during debugging and weakest during disputes. It excels at helping engineers reason forward. It struggles when asked to stand still and testify.

This is where the automotive world offers a sharp contrast.

AMG could not operate as an isolated engineering island, even when working independently. Any serious collaboration with Cadillac or GM would begin not with power targets or suspension rates, but with diagnostics. Before you can argue about tuning, you have to agree on what the car is allowed to say about itself.

By the early 1990s, the industry had already learned this lesson. Diagnostics were standardized not for convenience, but for accountability. On-board diagnostics were designed so independent tools could interrogate the same truth surfaces the manufacturer relied on. CAN bus reporting existed so that multiple control units could speak in a shared language, ordered against a common clock.

Even the diagnostic scanner you borrow from an auto parts store participates in this agreement.

It does not invent truth. It queries it.

That scanner assumes that when it asks, "what happened at this moment," the answer will be ordered, timestamped, and comparable across subsystems. It assumes that engine events, transmission events, emissions events, and fault codes all agree on what a second means. It assumes that causality has already been enforced before it arrives.

This is not accidental. It is the result of negotiated standards.

AMG would insist on this alignment before touching performance. They would not accept a Fleetwood whose diagnostics told incompatible stories depending on which module you asked. They would not tune a car whose fault data could not be cross-correlated cleanly. They would not accept instrumentation that required interpretation before trust.

Agreement on diagnostics is agreement on authority.

This is the critical difference between reconstruction and governance. Automotive diagnostics are designed to be authoritative at the moment of observation. They are not reconstructed later by inference. They are anchored to a shared clock tick, enforced on the bus, and made accessible through standardized interfaces.

The system enforces order first. Diagnostics confirm it second.

Modern observability systems often reverse this order. They allow disagreement at the source, then attempt to reconcile it later. Automotive diagnostics never made that concession. They recognized early that if subsystems could not agree on time, they could not agree on responsibility.

AMG's collaboration with GM would depend on this shared diagnostic substrate. Not because of bureaucracy, but because without it, no tuning decision could be defended. Performance without shared diagnostics is exhibition. Engineering without shared time is speculation.

Modern software systems quietly abandoned this discipline.

They standardized interfaces. They standardized protocols. They standardized APIs.

But they allowed time to fragment.

And then they built telemetry to cope with the consequences.

This does not make telemetry dishonest. It makes it compensatory. Telemetry helps you live with broken time. It does not fix it.

Tracing can reassert order. It can report local duration. It cannot reassert authoritative time.

If you need correlation to know what happened, time has already failed.

That is the line between explanation and evidence.

Telemetry can help you understand a system. Shared diagnostics anchored to sovereign time can help a system testify.

For AMG, that distinction is non-negotiable. For systems that must answer under scrutiny, it should be non-negotiable as well.

Telemetry is reconstruction, not truth.

It is important to distinguish between authority and diagnostic capability.

Modern observability systems represent a remarkable technical achievement. The ability to collect, normalize, correlate, and analyze telemetry across disparate runtimes, languages, schedulers, and network boundaries would have been inconceivable a generation ago. Frameworks such as OpenTelemetry have enabled engineering teams to reason across systems that do not share clocks, identities, or execution context.

Where time is inconsistent or irreconcilable, these systems provide something invaluable: a relative ordering of dependent events that cannot be reconstructed by any other means. Even when absolute timestamps cannot be trusted, causal dependency graphs can still be inferred. That inference enables diagnosis, triage, and repair in situations where direct evidence has already been lost.

This capability should not be minimized. It is often the only way to make progress once structural authority has already been compromised.

But diagnostic reconstruction is not the same thing as testimony. A relative sequence is not a shared frame. Correlation is not jurisdiction. These systems explain how failure unfolded; they do not establish an authoritative account of what actually happened. Their value is instrumental, not evidentiary.

Telemetry excels precisely where authority has already failed. It allows engineers to work backward through ambiguity. It does not remove the ambiguity itself.

Legal, Forensic, and Executive Consequences

Untrusted time does not create confusion. It creates exposure.

Most discussions of clock drift end in engineering discomfort: harder debugging, longer incidents, noisier dashboards. That framing understates the cost. When time is not authoritative, the failure is not technical—it is institutional.

Responsibility depends on sequence. Liability depends on sequence. Authority depends on sequence.

Without a defensible answer to when, every other answer becomes provisional.

Audit trails are the first casualty.

An audit trail is not a collection of events. It is an ordered record that asserts continuity: this happened, then this happened, then this happened. Auditors are not looking for volume; they are looking for coherence. They assume that timestamps mean the same thing across records, systems, and boundaries. They assume that order is enforced, not inferred.

When logs disagree, that assumption collapses.

Two records that cannot be reconciled temporally do not merely create ambiguity—they create optionality. Optionality is dangerous. It allows opposing interpretations to coexist. It allows intent to be questioned. It allows diligence to be doubted.

From a legal perspective, disagreement is not neutral. It widens the surface of liability. If the system cannot assert a single timeline, opposing counsel will. And their timeline will not be charitable.

Chain of custody suffers next.

Chain of custody exists to preserve trust across transitions. Evidence moves from system to system, person to person, boundary to boundary. At each handoff, time anchors responsibility: who had control, for how long, and before or after which event.

If timestamps drift, custody blurs.

An event that appears to occur before it is recorded undermines provenance. A record that appears modified before it is created undermines integrity. Even if the underlying data is correct, the timeline invites challenge. And challenge is often enough.

This is why forensic disciplines obsess over time synchronization. Not because time is interesting, but because it is decisive. A single inconsistency can discredit an otherwise sound body of evidence.

Regulatory reporting raises the stakes further.

Regulators do not accept narrative. They accept records. Those records must align not only internally, but with external clocks: filing deadlines, reporting windows, statutory thresholds. Time is jurisdictional. Miss it, and intent becomes irrelevant.

When systems cannot agree on time, compliance becomes probabilistic. Reports are generated from data whose order is assumed rather than enforced. Cutoffs are approximated. Windows blur. Organizations begin to rely on "close enough" without realizing that regulators do not.

This is where well-functioning systems begin to fail on paper.

The system may operate correctly. The reports may be accurate in aggregate. The timestamps may still betray you.

Incident reconstruction exposes the problem publicly.

When something goes wrong—an outage, a breach, a safety event—the first demand is explanation. What happened? When did it start? When did we know? When did we act?

These are not technical questions. They are accountability questions.

If logs disagree, the incident timeline becomes negotiable. If metrics contradict each other, causality becomes debatable. If traces reconstruct order but not time, duration becomes speculative. Every gap invites suspicion.

At that point, engineering explanations give way to interpretation. Postmortems become narrative exercises. Confidence erodes—not because the team acted incorrectly, but because it cannot prove that it acted correctly.

This is where executive responsibility becomes personal.

Executives do not testify about systems. They testify about decisions.

Those decisions are anchored in time. What did you know at 10:14? What information was available at 10:18? Why was action taken at 10:22 instead of 10:12?

Without authoritative time, those answers become soft. And soft answers harden quickly under questioning.

If logs disagree, liability expands. If timestamps drift, evidence weakens. If causality is unclear, accountability evaporates.

This is not hypothetical. It is why organizations lose cases they should win. It is why settlements occur without admissions. It is why "we did the right thing" fails when the record cannot support it.

The AMG Fleetwood would never tolerate this ambiguity.

AMG's insistence on shared diagnostics and enforced clocks is not about convenience. It is about defensibility. When the car is evaluated—by regulators, insurers, or engineers—the data must stand on its own. The system must be able to testify without interpretation.

That discipline scales.

Systems that enforce time can explain themselves under pressure. Systems that reconstruct time can explain themselves only when everyone agrees to be reasonable. Authority does not depend on reasonableness. It depends on proof.

This is why untrusted time collapses responsibility. It shifts systems from evidence to persuasion, from testimony to storytelling. It forces leaders to defend decisions with confidence instead of records.

Confidence is fragile. Records endure.

If you cannot prove when something happened, you cannot prove that it happened. And if you cannot prove that, you cannot reliably assign responsibility, defend action, or assert authority.

Time is not a technical detail. It is the foundation of accountability.

This is the cost modern systems quietly accept when they allow time to drift, to fragment, or to be inherited without enforcement. They gain flexibility. They gain scale. And they mortgage their ability to stand still under scrutiny.

When everything is moving fast, that trade feels invisible.

Until someone asks you to stop—and explain.

Why "When" Matters More Than "What"

At the end of every failure, every dispute, every investigation, the questions narrow. The dashboards are closed. The graphs stop moving. The volume recedes. What remains is not how much or how fast, but sequence.

What happened when.

"What happened" by itself is an incomplete question. It invites speculation. It allows stories to compete. Two people can agree entirely on what occurred and still disagree fundamentally on responsibility if they cannot agree on when it occurred. At that point, explanation becomes persuasion.

Authority does not live in outcomes. It lives in order.

Order is what turns events into evidence. It is what allows cause to precede effect, decision to precede consequence, and responsibility to be anchored to a moment rather than inferred from a result. Without order, systems still act. People still decide. But accountability floats.

This is why time is not a supporting detail. It is the coordinate that makes everything else legible.

In the Fleetwood, performance is meaningless without sequence. Saying the car reached a certain speed tells you nothing until you know when it did so relative to braking, steering input, suspension load, and road conditions. The same speed reached before a maneuver and after a maneuver describe two entirely different vehicles. One is controlled. The other is lucky.

AMG understands this instinctively. They do not evaluate performance as a snapshot. They evaluate it as a progression. They care about transitions, not just endpoints. They ask how the car arrived where it did, not merely where it ended up. That is why they insist on trusted instrumentation before trusted performance.

A car that performs without trustworthy gauges is not impressive. It is indefensible.

There is one environment where this problem is often invisible: the monolith.

When everything runs on a single host—classic LAMP or WAMP stacks, large single-process applications—the system lives under the governance of one clock. That clock may drift. It may be wrong relative to the outside world. But it is wrong consistently. Every component agrees on "now," even if "now" is inaccurate.

In that world, sequence is preserved by accident.

Logs line up because they share a clock. Metrics correlate because they are measured against the same second. Incident reconstruction is straightforward because ordering is unambiguous. Causality survives not because time is authoritative, but because it is singular.

This is why monoliths often feel easier to reason about. Not because they are simpler systems, but because they still inhabit one temporal jurisdiction.

Modern systems rarely remain confined to that comfort.

Once an application fans out across dozens, or hundreds, of hosts, the illusion collapses. Each host brings its own clock. Each clock drifts according to its own enforcement policy. Each subsystem reports events against a local notion of time that may or may not agree with its neighbors.

At that scale, attempting to reconcile a transaction using logs alone becomes an exercise in futility. You are no longer drawing a timeline. You are assembling a hairline. And anyone who has lost their hair knows how thin that line becomes.

This is the point at which correlation IDs, tracing, and reconstruction become mandatory. Not because systems became more sophisticated, but because time stopped being shared. Sequence must now be inferred. Order must be rebuilt. Causality must be argued for rather than observed.

The monolith hid the problem. Distribution exposed it.

This is not an argument against distribution. It is an argument for honesty. Once one system no longer means one clock, "when" stops being self-evident. Without deliberate enforcement, it becomes probabilistic.

And probability is not authority.

Authority requires ordering. Ordering requires time sovereignty.

A system that cannot assert "this happened before that" cannot reliably assign responsibility, defend decisions, or withstand scrutiny. It can operate. It can scale. It can even appear stable. But when asked to stand still and explain itself, it falters.

This is why so many modern systems feel simultaneously observable and unprovable. They can tell you a great deal about themselves. They can explain flows, dependencies, and outcomes. But they cannot testify. They narrate.

That distinction is subtle until it matters.

When stakes are low, narrative is enough. When everyone agrees to be reasonable, inference works. When pressure is absent, plausibility suffices. But authority is not built for ideal conditions. It is built for moments when agreement breaks down.

Under pressure, only order holds.

AMG would never trust a performance claim that could not be defended by its instruments. They would never accept diagnostics that required interpretation before trust. They would never tune a car whose data could not be cross-examined cleanly.

That discipline is not nostalgia. It is governance.

Modern systems often invert this priority. They trust results first and instruments second. They assume time will behave because it usually does. They rely on telemetry to explain discrepancies when it doesn't. Over time, this creates comfort with approximation.

Approximation works until responsibility is required.

At that moment, the system must answer questions it was never designed to answer: What did we know at the time? When did we know it? What changed, and in what order? Why did we act when we did?

These are not engineering questions. They are authority questions. And they cannot be answered retroactively if order was not enforced prospectively.

This is the quiet cost of treating time as metadata instead of structure. Once time is demoted, everything that depends on it becomes conditional. Decisions become defensible only if others accept the story being told.

Authority does not require consensus. It requires proof.

Proof requires order. Order requires time sovereignty.

That is why "when" matters more than "what." Not because facts are unimportant, but because facts without order cannot be anchored. They drift. They accumulate. They contradict. They invite reinterpretation.

At the end of this chapter, the doctrine is simple and unavoidable:

If you cannot prove when something happened, you cannot prove that it happened.

Everything else—performance, scale, resilience, observability—rests on that foundation.

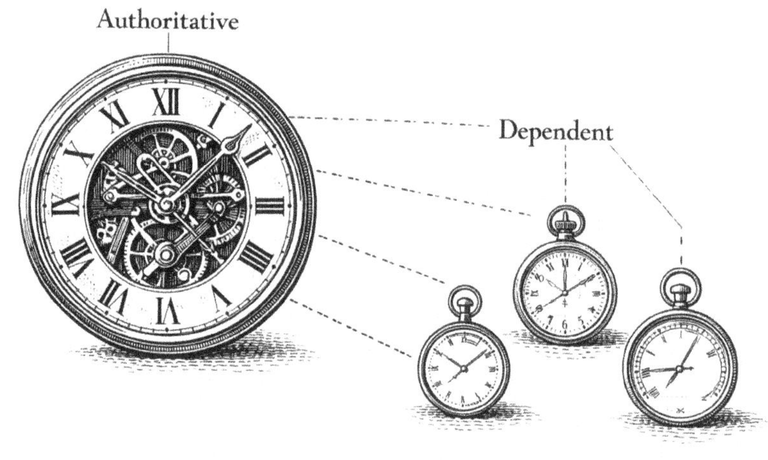

Time is the reference frame of causality.

5

Feedback Is Not Governance

Steering Is Direction, Not Authority

Steering is the most visible control surface on a car. It is the one the driver touches constantly. It is also the one most likely to be misunderstood.

Steering does not hold geometry. It does not absorb force. It does not define limits.

Steering points. That is all it does.

Every other assumption about control is borrowed.

When you turn a steering wheel, you are not commanding the car to obey. You are requesting a change in direction, and that request is honored only if the rest of the system is capable of honoring it. The tires must be loaded correctly. The suspension must preserve geometry. The frame must remain rigid. The brakes must be able to refuse speed. Time must be ordered. Authority must already exist.

Steering assumes all of that.

Without those preconditions, steering becomes theater—motion without consequence, input without effect, effort without authority.

This distinction is easy to miss because steering feels like control. It is immediate. It responds. It moves. But responsiveness is not governance. A system can respond continuously and still be directionless. It can feel alive while drifting.

That is why steering without a rigid frame is dangerous. When the frame flexes, steering inputs no longer correspond cleanly to outcomes. The wheel turns, but the car hesitates, lags, or overshoots. Geometry shifts under load. The driver compensates. Corrections stack. Oscillation begins.

The steering wheel did not fail. The assumptions beneath it did.

The same is true when brakes are weak or inconsistent. Steering can point the car toward safety, but it cannot remove energy. Without the ability to refuse speed, direction becomes academic. You may know where you want to go. You simply cannot stop in time to get there.

And suspension matters just as much. Steering relies on honest mediation between road and structure. If the suspension lies—by smoothing too much or breaking under load—steering feedback becomes untrustworthy. The wheel may feel calm while the tires are losing grip. Or it may feel nervous while the chassis is stable. In both cases, the driver is misled.

Steering assumes honesty elsewhere.

This is why AMG's steering philosophy diverged so sharply from Cadillac's during the period we are examining. Cadillac optimized for isolation. Overboosted steering reduced effort. Road feel was filtered aggressively. Center was vague, not because the engineers were incompetent, but because the brand promised effortlessness. The steering wheel was meant to reassure, not inform.

AMG made the opposite choice. Steering weight was deliberate. Resistance was intentional. Effort was part of the signal. The wheel did not exist to soothe the driver. It existed to tell the truth. Feedback was information, not control. The driver was expected to work, because work was how the system communicated limits.

This was not a preference. It was a governance decision.

AMG understood that steering should never be asked to compensate for deficiencies elsewhere. If the frame flexed, fix the frame. If the brakes faded, fix the brakes. If the suspension lied, retune the suspension. Steering was not a place to hide problems. It was a place to reveal them.

That philosophy maps cleanly to systems design.

Feedback loops are the steering wheels of software systems. Metrics, dashboards, alerts, traces—all of these are steering inputs. They tell you where the system is pointing. They tell you what is happening. They feel like control because they are immediate and visible.

But feedback loops do not create authority.

They assume it.

A dashboard does not decide whether a system is allowed to change. An alert does not determine whether an action should be taken. A metric does not enforce limits. They point. They inform. They request attention.

The authority to act must exist elsewhere.

This is where modern systems often confuse responsiveness with control. They build elaborate feedback mechanisms and then allow those mechanisms to drive behavior directly. Autoscaling responds to load. Alerting triggers reconfiguration. Optimization routines adjust parameters continuously. The system appears adaptive, alive, intelligent.

But without governance, this is steering without a frame.

Feedback loops are not control loops. A control loop has boundaries. It has invariants. It has refusal built in. It knows what it is not allowed to do, even when pressure increases. Feedback loops, by contrast, only know what they see. They react to signals, not to intent.

Metrics inform; they do not decide. Dashboards point; they do not govern.

When systems treat feedback as authority, they begin to drift. Local adjustments accumulate. Corrections stack. Oscillation replaces direction. The system becomes busy without becoming correct.

This is why teams often feel like they are constantly turning the wheel. Every metric change prompts action. Every alert demands response. Every dashboard movement invites adjustment. The system is always responding, always correcting, always in motion.

And yet it goes nowhere straight.

The steering wheel is being turned constantly because the car has lost its geometry.

Authority lives in what does not move. It lives in invariants: identity boundaries, time sovereignty, structural limits, refusal paths. Steering can exercise that authority, but it cannot create it. Without invariants, steering amplifies noise.

AMG never asked steering to do more than it could. They did not tune around a weak frame or mask brake fade with isolation. They insisted that authority be structural, so that steering could remain honest. The wheel pointed because the car could obey.

Systems should make the same demand.

Feedback is indispensable. Without it, systems are blind. But feedback is not governance. It does not decide what is allowed to change, when change must stop, or who has the right to override automation. Those decisions require authority, not information.

This is the core distinction of the chapter, and it must be made early and mechanically.

Feedback cannot create authority. It can only exercise authority that already exists.

Everything that follows in this chapter—runaway feedback, closed loops without judgment, the separation of governance as a system—depends on that truth. Steering is direction, not authority. When systems forget that, they do not fail dramatically. They wander convincingly.

And wandering, when it feels like control, is the most dangerous failure mode of all.

When Feedback Becomes Self-Reinforcing

Feedback is supposed to stabilize a system. It observes deviation and proposes correction. When governed, it is a powerful ally. When ungoverned, it becomes an accelerant.

The danger rarely announces itself as failure. It announces itself as responsiveness.

In the automotive world, the warning signs are familiar. Over-assisted steering can oscillate. The wheel becomes light, corrections stack, and the driver chases a center that no longer exists. Speed-sensitive steering without frame discipline feels clever at first—easy at low speed, weighted at high speed—until load changes expose the truth. The system is reacting faster than the structure can support. The most vivid example is death wobble: a small disturbance excites a loop, the loop feeds itself, and the vehicle shakes violently until energy is forcibly removed.

Nothing "broke" in the moment. The system simply amplified its own corrections.

The steering was not wrong. The absence of refusal was.

Feedback loops, by definition, feed outputs back into inputs. That is their power. But without boundaries, that power compounds. Each correction becomes the cause of the next deviation. Effort increases. Motion intensifies. Stability vanishes.

The same pattern plays out in software systems.

Positive feedback loops are easy to create and hard to recognize, especially when they wear the mask of optimization. A metric moves. A system responds. The response moves the metric further. The system responds again. What looks like adaptation is actually acceleration.

Local optimization is the most common trigger. A service sees increased latency and scales out. Scaling increases contention elsewhere. That contention produces new signals. The system responds locally again. Each component behaves "correctly" within its own narrow frame. Globally, the system drifts into incoherence.

This is how global truth is poisoned by local success.

Metric chasing is a particularly pernicious form of this failure. Teams select a number, declare it meaningful, and then allow that number to drive behavior. Latency goes up, so capacity is added. Error rates spike, so retries increase. Throughput dips, so parallelism is expanded. Each response improves the metric in isolation while degrading the system as a whole.

The metric improves. The truth worsens.

Alert-driven reconfiguration follows the same pattern. Alerts are designed to inform humans, but in many modern systems they have been wired directly into automation. An alert fires, a script runs. A threshold is crossed, a change is made. The loop tightens. Humans are removed not because the system is more intelligent, but because the system is faster.

Speed is mistaken for judgment.

Autoscaling is the canonical example. At modest scale, autoscaling is a kindness. It absorbs burst, preserves availability, and buys time. At large scale, ungoverned autoscaling becomes reflex. Load appears, capacity expands. The expansion masks inefficiency. Inefficiency attracts more load. The system grows to fill the resource pool available to it. Costs rise. Complexity rises. Failure modes multiply.

The system is responding constantly. It is not deciding anything.

This is steering without a frame.

There is a simpler way to see why ungoverned feedback never converges.

A carnival balloon animal obeys the same rules as an ungoverned system. The total volume is fixed. The material is elastic. Pressure redistributes rather than disappears. Grab the tail and the front legs grow. Squeeze the legs and the nose inflates. Pinch the nose and the ears swell.

Nothing intelligent is happening. Nothing is being corrected. Energy is merely being displaced.

The balloon does not know what shape it is supposed to be.

Every intervention creates a new distortion somewhere else. Each correction feels locally effective and globally wrong. The system responds instantly, convincingly, and endlessly—because there is no invariant telling it where change must stop.

This is feedback without refusal.

Modern systems behave the same way when optimization is allowed without governance. Load is relieved in one place and reappears in another. Latency is reduced in one service and emerges downstream. Error rates fall locally while retrics amplify pressure globally. Autoscaling absorbs stress in one tier and pushes it into cost, contention, or complexity elsewhere.

The system is not improving. It is redistributing strain.

Each adjustment looks like progress because something got better. What is rarely measured is what got worse as a consequence. Without a governing boundary—without a segment that is not allowed to change—the system becomes a balloon endlessly reshaped by well-meaning hands.

This is why local optimization poisons global truth. Every team squeezes where it hurts. Every metric rewards relief. The system obliges by swelling somewhere else. Eventually no one recognizes the original shape, but everyone feels busy fixing it.

The balloon never stabilizes because stabilization was never encoded.

What the balloon analogy reveals is redistribution without judgment. What it also reveals is the absence of an anchor.

Governance requires a business anchor.

Governance is not the act of preventing change. It is the act of deciding which changes matter. Many things can be adjusted in a system. Very few should be. Governance exists to answer a narrow question that feedback loops cannot ask on their own:

Which one or two changes produce the greatest business impact, and everything else must yield to that priority?

Feedback loops have no concept of business value. They operate on immediacy, not importance. They respond to pressure, not consequence. A number moves, so a response is triggered. Velocity is preserved. Value is assumed.

This is why ungoverned automation trades continuous motion for diminishing returns. Systems become very good at changing themselves while becoming progressively worse at changing the right thing. Every metric becomes eligible for optimization because no metric is designated as dominant. The system moves constantly because nothing is allowed to be decisive.

Business impact is what stops that motion.

An anchor does not move simply because pressure exists. It defines what success actually means and forces all other adjustments to subordinate themselves to that definition. Without it, automation will always optimize the wrong thing perfectly.

This is the failure mode automated feedback loops exhibit most reliably. They are excellent at preserving velocity. They are terrible at selecting priorities. They do not know which changes reduce risk, protect revenue, or preserve trust. They know only that a signal appeared and that a response is possible.

That is not judgment. It is reflex.

This is where governance asserts itself. Governance decides which dimensions are allowed to move and which must remain invariant. It defines refusal points that automation cannot override. It says, "This outcome matters more than responsiveness, more than utilization, more than elegance."

Software Performance Risk Management exists precisely to address this gap. Its core premise is not that systems lack data, but that they lack direction. Knowing what to change for the greatest business impact is a governance decision, not a feedback decision. Automation can execute that decision. It cannot originate it.

AMG would recognize this instantly. They would never allow every adjustable parameter to move simply because it could. They would decide what mattered first—composure, predictability, braking authority, stability under load—and then tune everything else around those invariants. Steering assist, damping, power delivery all serve the same question: does this make the car better at its purpose?

Not faster in isolation. Not more responsive in theory. Better at the job it exists to do.

Systems that lack this anchor will always prefer velocity over value. They will celebrate constant adjustment while quietly eroding the outcomes the business actually depends on. The feedback loops will work. The dashboards will glow. The organization will feel busy.

And the system will drift.

Drift is the natural outcome of feedback without governance. It does not crash the system. It rotates it away from intent while preserving the illusion of control.

In vehicles, death wobble is resolved by adding refusal: stiffer components, better damping, tighter tolerances. The goal is not to eliminate feedback, but to bound it. The steering wheel still turns. The system still responds. But there are limits that cannot be exceeded, regardless of input.

Systems require the same humility.

A feedback loop that cannot be told "no more change here" will never stop on its own. It will continue correcting long after correction has ceased to be useful. It will amplify noise because noise looks like signal when nothing is anchored.

This is why the critical line must be stated plainly and remembered when systems feel most alive:

Feedback without refusal is acceleration, not correction.

Everything that follows in this chapter—closed loops without judgment, governance as a separate system, the illusion of responsiveness—rests on this failure mode. Steering is not authority. Feedback is not governance. When systems confuse the two, they do not collapse dramatically.

They wander convincingly.

Closed Loops Without Judgment

Automation is seductive because it feels decisive. A loop closes, a signal is acted upon, a deviation is corrected. Something happens. Stability appears. Charts flatten. Noise subsides. The system looks calm.

But calm is not correctness.

In mechanical systems, this distinction is well understood. A PID controller can stabilize almost anything given enough tuning. It can damp oscillation, reduce error, and hold a value within tolerance. That does not mean the system is doing the right thing. It means the system is doing a consistent thing.

Stability is a property of motion. Correctness is a property of intent.

A Proportional-Integral-Derivative (PID) controller does not know why it is stabilizing a value. It does not know whether the target is appropriate, whether the environment has changed, or whether the goal itself is now harmful. It minimizes error against a reference. That is all.

This is why engineers never confuse control loops with judgment. Loops execute. Humans decide.

Local maxima are the classic trap. A system finds a configuration that appears optimal within its immediate feedback horizon and settles there. From the inside, everything looks stable. From the outside, the system is stuck. Escaping that local optimum requires violating the loop's assumptions—introducing disturbance, redefining the goal, or overriding the controller entirely.

A loop cannot decide to leave a local maximum. Only judgment can do that.

AMG understands this instinctively. They use automation where automation belongs, but they never allow software to "discover" geometry. They do not let algorithms decide what suspension travel should mean, what steering effort should communicate, or where limits must exist. Humans define those bounds first. Automation operates inside them.

Loops are tuned by hand because the consequences matter.

AMG's philosophy is not anti-automation. It is anti-delegation of intent. They recognize that loops are excellent servants and terrible governors. Stability without purpose is just equilibrium at the wrong place.

Modern systems struggle with this distinction because automation has moved closer to decision-making than ever before. Algorithms are no longer confined to execution. They are increasingly asked to infer goals from data. Control planes adjust systems they do not understand. Machine learning models optimize against historical truth that is already compromised.

The loop closes. The chart improves. The system drifts.

Algorithms optimizing corrupted truth are especially dangerous because they amplify invisibly. If the data reflects bias, error, or past compromise, the optimization process will reinforce it. The model does not ask whether the data should be trusted. It assumes it should be optimized.

This is feedback with a halo.

The most troubling version appears when systems train on their own outputs. Recommendation engines reinforce what they previously recommended. Fraud models learn from labels they influenced. Ranking systems adapt to behavior they shaped. The loop tightens. Error compounds. Bias hardens into structure.

The system is stable. It is also wrong.

This is the defining danger of closed loops without judgment. They do not fail loudly. They converge confidently on the wrong answer.

The problem is not intelligence. It is authority.

A loop that cannot refuse its own input will amplify error indefinitely. It has no mechanism for doubt. No concept of override. No place where intent can reassert itself. Every output becomes the next input. Every assumption calcifies into fact.

Judgment requires a break in the loop.

It requires someone—or something—outside the feedback cycle to ask questions the loop cannot ask:

- Is this still the right goal?
- Are we optimizing the right dimension?
- What changed that the model cannot see?
- What must not be allowed to move?

Automation cannot answer these questions because they are not computational. They are contextual. They require understanding consequence, not minimizing error.

A contemporary example makes this painfully clear.

In the days leading up to the Super Bowl, systems behave exactly as designed. Traffic is steady. Load is predictable. Clusters are right-sized. Autoscaling policies are tuned for efficiency. From the system's perspective, nothing unusual is happening.

Then the commercial airs.

Traffic spikes by orders of magnitude in seconds. The control plane reacts instantly. New instances are scheduled. Containers spin up. Dependencies warm. The system is busy, responsive, alive.

By the time capacity arrives, the moment is gone.

From the loop's perspective, everything worked. From the business's perspective, the opportunity evaporated.

This is not a failure of automation. It is automation doing exactly what it was told—optimizing for historical truth. The system assumed tomorrow would look like yesterday. It had no knowledge of intent, no awareness of consequence, no understanding of business time.

The question that mattered most was never asked:

Is this a moment where cost efficiency should yield to certainty?

That question does not belong to a control loop. It belongs to governance.

In well-governed systems, humans intervene before known load arrives. Capacity is pre-staged. Baselines are deliberately violated. Resources are allocated not because metrics demand them, but because intent requires them. Authority acts early so automation does not have to act late.

Often, that authority is delegated intelligently. Content delivery networks absorb the surge. Static assets are made publicly cacheable. Private content is pushed to the edge where possible. Origin systems are protected precisely because someone decided, in advance, what must not fail.

When this is done well, the control loops never see the spike. The system remains calm not because it adapted quickly, but because it was never asked to.

Left alone, automation will always optimize for the present. It cannot sacrifice efficiency today for certainty tomorrow unless explicitly instructed to do so. It does not know which moments matter more than others. It does not know when being "overprovisioned" is not waste, but insurance.

The human in the governance loop wins not by reacting faster, but by acting earlier.

This is the essential boundary between automation and authority. Closed loops are invaluable for maintaining equilibrium. They are useless for recognizing inflection points that exist outside their historical horizon.

AMG would never rely on a control loop to prepare a car for an event it knew was coming. They would not wait for temperatures to rise before adding cooling. They would not let suspension loads dictate setup after the first lap. They would stage the system deliberately, based on knowledge the car itself does not possess.

Systems deserve the same respect.

Closed loops execute policy. Judgment defines purpose.

Governance exists to reintroduce friction where loops would otherwise accelerate. It is the deliberate act of saying, "This stops here." Not because the system cannot continue, but because it should not.

That refusal is not failure. It is judgment made visible.

A loop that cannot be told "no" will always say "yes" to itself. It will optimize until it collapses into a local maximum and call that success. Only authority—external, intentional, anchored—can break that spell.

This is the boundary this chapter is establishing. Automation executes. Feedback informs. Loops stabilize. Governance decides.

When systems forget this order, they do not implode. They ossify. They become perfectly stable machines doing the wrong thing forever.

And that, more than any outage, is the failure that matters.

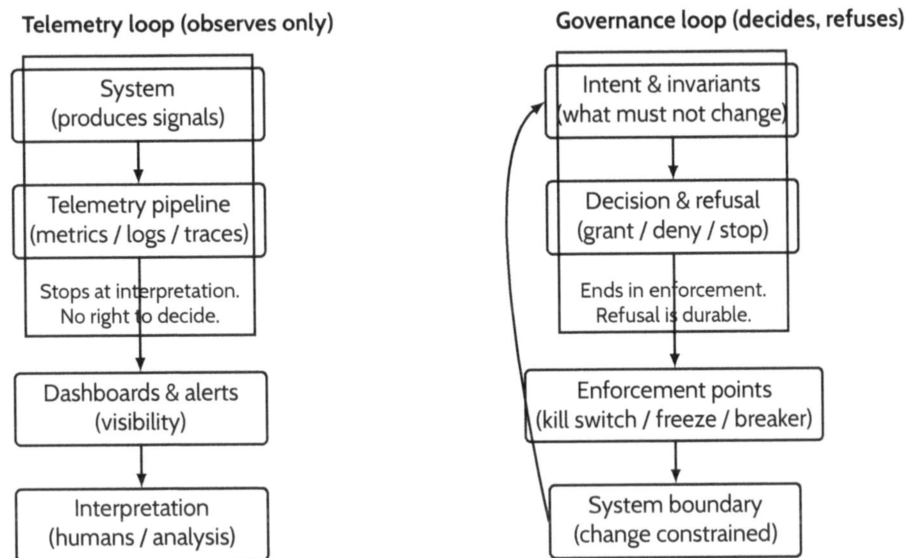

Figure 5.1: Telemetry describes the system. Governance constrains it. Visibility is not jurisdiction: observation cannot become authority by repetition.

Governance Is a Separate System

Governance is often mistaken for slowness.

It is described as bureaucracy, friction, or hesitation. In reality, governance is none of those things. Governance is decision made durable. It is the mechanism by which intent survives pressure, scale, and time.

Feedback reacts. Governance refuses.

This distinction matters because feedback and telemetry are visible, while governance is often silent. Dashboards glow. Alerts fire. Traces animate. Governance, when it is working, does very little at all. It sits in the background, enforcing boundaries that are rarely tested—until they are.

The mistake modern systems make is assuming that because feedback is active, it must also be authoritative. It is not. Feedback can propose. Telemetry can inform. Neither can decide what is allowed to change.

That decision belongs to governance, and governance must exist as a separate system.

In mechanical design, this separation is explicit. Alignment specifications are not suggestions. They are invariants. Torque values are not advisory. They are enforced. Safety margins are not tuned dynamically by sensors. They are defined by humans who understand consequence.

And when those values are set, they are signed off by someone whose name appears on the drawing.

The system does not get to renegotiate them under load.

AMG would never allow a feedback loop to redefine alignment specs mid-corner. They would never let a control system adjust torque limits because it "felt like" more power was available. Those decisions are made upstream, deliberately, and rarely. Automation executes within those bounds. It does not redraw them.

This is governance.

Governance decides what cannot change. It defines what must remain invariant even when conditions deteriorate. It establishes who has the authority to override automation, and under what circumstances. It owns rollback, not as an emergency reflex, but as a designed capability.

Feedback does not do this. Telemetry cannot do this. Control loops cannot do this.

They were never meant to.

In software systems, this separation is often blurred because the artifacts of feedback are so compelling. Metrics update in real time. Alerts demand attention. Control planes adjust continuously. The system appears alive. Governance, by contrast, looks inert. It lives in documents, policies, runbooks, and design decisions made long before anything goes wrong.

Because it is quiet, it is often undervalued.

Until it is absent.

Change control is governance. Not because it slows things down, but because it defines when speed must yield to safety. Freeze windows are governance. They are declarations that certain periods matter more than velocity. Circuit breakers are governance made executable. They refuse progress when continuation would cause harm. Kill switches are governance distilled to its simplest form: the right to stop.

These are not feedback mechanisms. They do not optimize. They do not react proportionally. They are binary by design because they exist to enforce boundaries, not explore them.

This is why governance must be designed as a system, not layered on as a policy. A policy without enforcement is a suggestion. Governance without execution is theater. If the system cannot refuse change when governance demands it, then governance does not exist.

Authority surfaces make this concrete.

Identity determines who is allowed to act. Time determines when actions are ordered and validated. Structure determines what can bear load. Governance binds these surfaces together and declares which aspects of them are immutable.

Identity without governance devolves into privilege sprawl. Time without governance devolves into narrative reconstruction. Structure without governance flexes until it lies.

Feedback can reveal problems on these surfaces. It cannot repair them.

This is why governance cannot live inside the feedback loop. A loop cannot override itself. A system cannot grant itself authority to stop. That authority must exist outside the loop, anchored in intent and consequence.

AMG's design process reflects this separation. There are decisions they revisit constantly—damping curves, assist levels, tuning parameters. And there are decisions they do not revisit at all—safety margins, load paths, structural limits. The latter are governed, not optimized. They are defined once, reviewed carefully, and then enforced relentlessly.

Automation works because governance exists.

Modern platforms often invert this relationship. They build elaborate feedback systems and assume governance will emerge organically from data. It does not. What emerges instead is oscillation. Teams adjust systems based on the loudest signals. Baselines shift. Invariants erode. The system remains active but loses its sense of direction.

This is how organizations end up with systems that are endlessly adjustable and fundamentally unreliable.

Governance is what prevents that outcome. It is not a reaction to failure. It is a precondition for trust.

Trust does not come from responsiveness. It comes from predictability under pressure. It comes from knowing that certain things will not change, regardless of what the metrics say in the moment. It comes from the confidence that someone, somewhere, has the authority to say, "This stops here."

That authority must be explicit.

Who can override automation? Who decides when rollback is mandatory? Who owns the consequences of change?

If these questions do not have clear answers, governance does not exist. Feedback may still function. Telemetry may still glow. But the system is operating without a governor.

This is why governance cannot be an afterthought. It must be designed as deliberately as any control plane. It must have interfaces, enforcement points, and escalation paths. It must be exercised occasionally to remain credible. And it must be respected even when it is inconvenient.

Especially when it is inconvenient.

Feedback reacts to what has already happened. Governance acts on what must not happen. Feedback can tell you that you are drifting. Governance decides whether drift is acceptable.

This is the line that separates automation from authority.

Feedback reacts. Governance refuses.

Everything else in this chapter flows from that distinction. Closed loops without judgment fail because they lack governance. Systems that feel busy but go nowhere lack governance. Platforms that optimize endlessly without delivering value lack governance.

AMG would never confuse steering, suspension, or control loops with authority. They know exactly where authority lives, and they protect it.

Systems must learn the same lesson.

Governance is not control. Governance is constraint with intent.

Steering Into Noise

Cadillac did not misunderstand steering. They understood their customer.

For decades, Cadillac's design mandate was explicit: the car should separate the driver from the road. Noise, vibration, and harshness were not sources of information to be preserved; they were intrusions to be removed. Luxury was defined as calm. Effortlessness was the promise. The road existed to be crossed, not conversed with.

That philosophy shaped everything downstream. Steering was boosted not to improve response, but to reduce work. Feedback was filtered not because it was inaccurate, but because it was unwanted. Straight-ahead feel was intentionally soft, because constant micro-correction was considered fatigue, not engagement. A Cadillac was meant to proceed forward without demanding attention.

This was not a failure of engineering. It was a coherent worldview.

AMG held a different one.

AMG believed that some noise is signal. That texture, load transfer, and surface variation are not distractions but information. Steering, in their view, was not meant to soothe. It was meant to communicate. Weight was intentional. Resistance carried meaning. Straight-ahead mattered precisely because deviation from it could be felt.

Neither model is wrong. But each carries a different risk when pushed too far.

In automotive terms, a "busy" steering feel is one where the wheel never quite settles. There is no calm center. The car requires constant correction, not because it is unstable, but because the system is too responsive. Every imperfection in the road, every small input, is amplified. The driver becomes occupied with maintaining direction rather than choosing it.

The car feels alive. It is also exhausting.

Nervous cars are rarely dangerous in isolation. They respond quickly. They change direction eagerly. But over time, the lack of a stable baseline erodes confidence. The driver stops trusting straight-ahead. Corrections stack. The system feels like it is always doing something, even when nothing needs to be done.

This is steering into noise.

Modern systems exhibit the same behavior when responsiveness is elevated to virtue without restraint. "Data-driven" organizations pride themselves on real-time awareness. Dashboards update continuously. Alerts fire instantly. Decisions are made on live data. The system is always adjusting.

And the baseline disappears.

When everything is actionable, nothing is stable. Teams react to every fluctuation. Minor deviations prompt changes. Changes introduce new deviations. Roadmaps are rewritten weekly. Priorities shift with metrics. The organization becomes very good at responding and very bad at deciding.

Noise masquerades as insight.

This is not because data is wrong. It is because data is abundant. Without governance, without declared invariants, the system treats every signal as equally important. Steering inputs are honored even when the car is already pointed straight.

The result is thrashing.

A quiet inversion often accelerates this failure, almost unnoticed.

In many modern systems, the volume of data collected about the system exceeds the volume of data the system actually delivers to users. Telemetry outgrows traffic. Logs outweigh payloads. Metrics streams eclipse business transactions. Observation becomes more data-intensive than operation.

At that point, the system is no longer primarily serving customers. It is serving its own narration.

This happens not because engineers are careless, but because the incentives align that way. The marginal cost of collection feels low. Storage is cheap. Pipelines scale. The perceived risk of missing something feels existential. And the person deciding what to collect is often not the person accountable for what it costs to store, process, and interpret it.

So the rational choice becomes: collect everything.

Just in case.

Telemetry systems rarely distinguish signal from noise at the point of collection. That decision is deferred—to analysis, to tooling, to dashboards, to humans downstream. But once the data exists, it asserts influence. It demands storage. It demands indexing. It demands explanation. It demands attention.

Quantity becomes authority.

When observation outgrows intent, responsiveness becomes compulsory. Dashboards fill. Alerts multiply. Teams feel obligated to react because the system is telling a story—regardless of whether that story matters.

The organization begins steering into noise because noise is now the loudest thing in the room.

This is not fundamentally a tooling problem. It is a governance problem.

No one decided which questions were worth answering. No one declared which signals mattered more than others. No one imposed refusal at the point of collection. The only governor was price—and price was abstracted away from decision-making authority.

So the system optimized for completeness instead of meaning.

Cadillac faced the same trade-off in mechanical form. They could have transmitted every vibration, every ripple, every imperfection directly to the driver. The road contains infinite data. They chose not to. They decided that most of it was noise relative to the experience they intended to deliver. Isolation was not denial. It was curation.

AMG made a different curation choice, but it was still curation. They did not transmit everything. They transmitted what mattered to control. Feedback was selective, shaped, and purposeful. Straight-ahead remained protected.

In both cases, authority existed upstream of sensation.

Systems that collect everything abdicate that authority. They postpone judgment and hope it will emerge later. It rarely does. What emerges instead is a culture of reaction, where the loudest signal wins not because it is important, but because it exists.

"How much is noise? How much is signal?" is not a statistical question. It is a governance question.

It cannot be answered by tooling. It must be answered by intent.

A system without collection boundaries will always feel busy. A team drowning in telemetry will always believe something requires action. The absence of refusal at the point of observation guarantees over-correction downstream.

Noise is not dangerous because it exists. It is dangerous because it feels actionable.

This is why governance must extend even to observation. Not every metric deserves to be collected. Not every event deserves to be logged. Not every trace deserves to be retained. Collection is a form of commitment, and commitment without judgment is drift.

AMG would never tune steering to feel busy. They would never accept a car that required constant correction to maintain direction. They would define center, protect it, and let the driver deviate deliberately.

Systems deserve the same respect.

Real-time data is valuable. Feedback is necessary. Responsiveness has its place. But when every signal becomes a steering input, the system loses its ability to travel straight. The driver becomes a servant to noise. The organization becomes reactive by default.

Noise does not disappear when you chase it. It multiplies.

The answer is not less data for its own sake. It is deliberate blindness in service of purpose. It is the confidence to say, "We will not collect this, because even if it changes, we will not act on it."

That is not ignorance. That is authority.

A system that steers into noise will always feel busy. A system with a compass will know when to hold course.

And holding course, when pressure rises, is the rarest form of control.

Authority Is the Right to Say "No Change"

Authority is often confused with activity. Systems that change frequently look alive. Organizations that respond instantly appear decisive. But neither motion nor speed is evidence of authority.

Authority is invariance under pressure.

It is the capacity to hold something still while everything around it demands movement. It is the ability to say, "Not this. Not now." And to have that refusal endure.

Feedback proposes. Authority disposes.

This distinction is the quiet center of the chapter. Everything before it—steering, loops, noise, governance—has been building toward this moment. Feedback shows you what happened. Telemetry explains how it happened. Automation can even suggest what might happen next. None of those things decide what must not happen.

Only authority does that.

In mechanical systems, this truth is unambiguous. Certain dimensions do not move, regardless of load. Alignment specs do not shift because the road is rough. Safety margins do not renegotiate because temperatures rise. Torque limits are not advisory when stress increases. These constraints exist precisely for the moments when pressure is highest.

If they moved under load, they would not be authority surfaces. They would be preferences.

AMG's design philosophy makes this explicit. Before anything is tuned, certain decisions are made once and then protected. Structural limits are defined. Load paths are fixed. Safety margins are set conservatively. Only after those invariants are declared does tuning begin.

What does not move comes first. Everything else is subordinate.

This is restraint as engineering, not as denial. AMG does not eliminate adjustability. They channel it. Damping curves can be refined. Assist levels can be tuned. Power delivery can be shaped. But none of those adjustments are allowed to compromise the invariants that define the car's character and safety.

Power follows restraint.

This is the opposite of how many modern systems are built. They begin with adjustability and hope invariants will emerge later. They expose dozens of levers, wire them to feedback, and allow automation to move them continuously. Over time, nothing remains fixed. The system becomes fluid in every dimension, responsive everywhere, authoritative nowhere.

When pressure arrives, there is nothing solid to push against.

Authority, by contrast, is visible most clearly when it refuses to move. When a system declines to adjust because doing so would violate intent. When it absorbs pressure without reconfiguring itself into something unrecognizable. When it protects the baseline even as metrics argue for change.

This is why authority feels slow only to those who mistake speed for correctness. In reality, authority is what allows systems to move quickly when it matters, because the boundaries are already known. Decisions are faster when fewer things are eligible to change.

Feedback is loud. Authority is quiet.

Feedback tells you that latency increased. Authority decides whether latency is allowed to increase. Feedback shows that load is rising. Authority decides whether cost efficiency yields to certainty. Feedback reports that users are behaving differently. Authority decides whether the system adapts or holds course.

Without that decision layer, feedback becomes tyranny by immediacy. The most recent signal wins. The loudest metric dominates. The system is pulled by whatever changed last.

This is not adaptability. It is drift.

Authority is what breaks that cycle. It says, "This dimension is closed." It removes entire classes of response from consideration. It enforces refusal not as an error condition, but as a design choice.

In systems, this shows up as decisions that feel uncomfortable at first. Freeze windows that prevent "quick fixes." Circuit breakers that stop progress abruptly. Kill switches that undo hours of automation in seconds. Identity boundaries that block action even when credentials exist. Time constraints that invalidate otherwise plausible data.

These are not failures of flexibility. They are expressions of authority.

They exist because someone decided, in advance, that certain outcomes are worse than delay, worse than inefficiency, worse than temporary pain. They encode values into the system that feedback alone cannot infer.

AMG encodes those values mechanically. Systems must encode them architecturally.

Authority is also what makes trust possible. Trust does not come from seeing everything. It comes from knowing that some things will not change arbitrarily. That the system will behave predictably under stress. That there is a line it will not cross, regardless of how compelling the data appears in the moment.

This is why authority cannot be emergent. It must be declared.

Who has the right to override automation? Who decides when rollback is mandatory? What conditions invalidate optimization entirely?

If these answers are ambiguous, authority does not exist. Feedback may still function. Telemetry may still glow. But the system is operating without a governor.

Cadillac and AMG again provide useful contrast. Cadillac achieved authority through isolation. They absorbed noise mechanically so the driver did not have to process it. AMG achieved authority through communication, but only because they protected the baseline. In both cases, authority existed upstream of sensation.

Neither brand relied on constant adjustment to define control.

Systems that chase every signal rarely realize they have lost authority until something irreversible happens. By then, invariants have already eroded. Refusal paths have been optimized away. Everything is adjustable, and nothing is dependable.

Authority is what prevents that erosion.

It is not reactive. It is anticipatory. It does not wait for feedback to justify itself. It exists precisely so that some feedback can be ignored.

That is the hardest lesson for modern systems to learn: not every signal deserves a response. Not every change deserves action. Not every improvement is worth making.

Authority decides which changes are allowed to matter.

This is the chapter's final claim, and it must land without drama because it is not controversial. It is simply true.

Feedback tells you what happened. Authority decides what happens next.

When systems invert that relationship, they become exquisitely aware and perpetually confused. When they honor it, they gain something rare: the ability to move decisively without losing themselves.

That is not rigidity. That is control earned through restraint.

And restraint, when it is deliberate, is the highest form of authority.

Governance exists where change is permitted or refused. Feedback exists where motion is optimized.

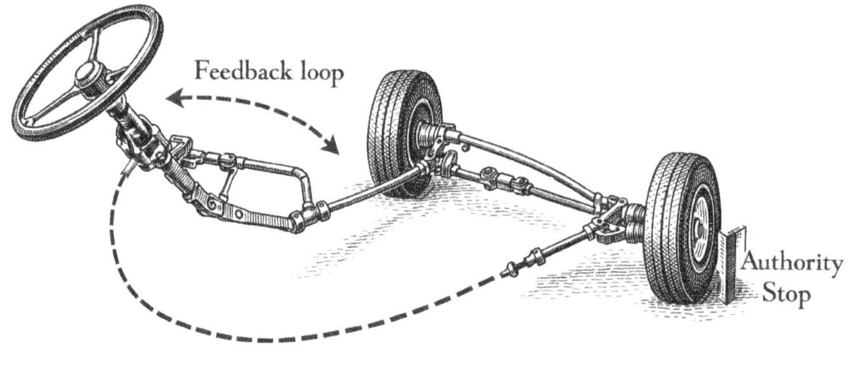

Feedback informs motion. Governance defines limits.

6

Chrome vs Monochrome

Performance is downstream from culture.

Not from specifications. Not from tooling. From culture. From what a system believes it is allowed to value, what it is expected to signal, and what it must never appear to compromise. In this case, the culture is unmistakably American—General Motors, Cadillac, and the particular definition of authority they spent a century teaching the public to recognize. This chapter is not about taste. It is about interface. About how authority is announced before it is exercised, and why a system that speaks the wrong visual language can be structurally sound and still be rejected. Only now—after frame, brakes, suspension, clocks, and governance—does appearance matter. Because now we understand what it must carry.

Appearance Is an Interface, Not Decoration

Before a system is touched, it is read.

Long before a driver feels steering weight or braking authority, they register stance, proportion, finish, and intent. Before an executive trusts a dashboard or an operator responds to an alert, they form a judgment about seriousness,

99

competence, and belonging. This happens instantly and involuntarily. Appearance is not decoration layered on top of function. It is the first interface a system presents to the world.

Interfaces set expectations. Expectations determine tolerance. Tolerance determines whether authority is granted or challenged.

This is why appearance cannot be dismissed as vanity. It is a control surface—one that shapes how much friction a system will encounter before it is allowed to operate. A system that looks authoritative is given room to prove itself. A system that looks ambiguous is interrogated immediately. A system that looks absent is dismissed, regardless of how correct it may be underneath.

There is no neutral design.

What people often describe as "clean," "minimal," or "objective" is simply a visual language associated with a particular culture. It carries assumptions about seriousness, restraint, and legitimacy that are learned, not universal. When that language is applied outside its cultural context, it does not disappear. It becomes illegible.

Automobiles make this obvious because they are read at a distance. You know whether a car intends to be authoritative before it moves. Wheelbase, overhangs, ride height, surface treatment, brightwork—these are not aesthetic flourishes. They are signals. They tell you whether the car expects to be deferred to, negotiated with, or ignored.

Cadillac understood this deeply. For much of its history, Cadillac did not aim to look fast. It aimed to look inevitable. Size was not aggression; it was reassurance. Chrome was not decoration; it was permanence made visible. The car announced itself as something meant to endure, to carry weight, to occupy space without apology.

This visual language did real work. It reduced friction before interaction. Other drivers yielded. Owners felt protected. Authority was granted before any mechanical claim was tested.

That is interface.

AMG's visual language performs the same function in a different culture. Monochrome finishes, restrained brightwork, and subdued badging signal internal confidence. They communicate that nothing needs to be proven. The car does not ask for attention because it assumes legitimacy. This works where authority is presumed and understatement is read as discipline.

But that reading is not universal.

When visual language crosses cultures without translation, correctness can read as absence. Restraint can read as cost-cutting. Discipline can read as indifference. A system can be structurally impeccable and still fail socially because it does not announce itself in a way the surrounding culture recognizes as authoritative.

This is why appearance must be treated as interface, not ornament.

The same failure appears constantly in software systems, most visibly in dashboards. A dashboard is not a neutral surface. It is an interface, and like any interface, it encodes assumptions about who is looking at it, what decisions they are empowered to make, and what language they trust.

Executives govern through business outcomes. Their authority is exercised through cost, margin, revenue, risk, and exposure. A dashboard that greets them with throughput charts, heap utilization, or trace waterfalls does not feel informative. It feels evasive. The signal does not map to the levers they are allowed to pull.

Operations teams inhabit a different reality. Capacity, saturation, queue depth, and throughput are not abstractions. They are control surfaces. An operations dashboard cluttered with financial rollups or executive summaries replaces clarity with noise. It presents context that cannot be acted upon and obscures what can.

SRE and DevOps engineers operate deeper still. They need traces, spans, dependency graphs, and transaction-level detail. Their work is forensic and temporal. A simplified, executive-style interface deprives them of the resolution required to preserve truth under load.

Each of these audiences is legitimate. Each of these interfaces is correct. They are not interchangeable.

When interfaces are mixed, authority collapses.

Executives confronted with engineering noise lose confidence and defer decisions they should make. Operators confronted with business abstractions feel blamed for outcomes they cannot directly control. Engineers confronted with sanitized summaries lose the ability to diagnose reality. No one is well served, because the interface no longer speaks a coherent language.

This is the systems equivalent of applying the wrong visual grammar to a flagship automobile.

A Fleetwood that looked like an AMG would confuse its audience before it moved. The car might be disciplined underneath, but the signals it sent would be misread as subtraction rather than restraint. Authority would be questioned before the engine ever turned over.

Interfaces grant permission. They tell the viewer, "This is your domain. These are your levers. This is what matters here." When that signal is muddled, decisions stall or misfire. Authority is either overreached or abdicated.

Legibility matters because it determines who feels empowered to act.

This is why appearance must come after authority is built but before authority is exercised. It cannot compensate for weak structure, but it can prevent unnecessary friction for strong systems. It is the difference between being tested and being trusted.

Cadillac's authority was meant to be seen. It was public-facing, declarative, and reassuring. It told the world, "This is substantial. This belongs here." AMG's authority was meant to be inferred. It spoke to an audience that already assumed competence and valued restraint. Each language worked precisely because it was culturally aligned.

Neither was universal. Both were intentional.

The mistake is not choosing the wrong aesthetic. The mistake is believing aesthetics are optional.

Appearance is how a system declares its intent. It tells the world whether it expects to be deferred to or debated. Whether it belongs or is visiting. Whether it understands the culture it operates within.

Only after frame, brakes, suspension, time, and governance are settled does appearance matter. But once it matters, it matters profoundly.

Because performance is downstream from culture. And culture is what decides whether authority is granted at all.

Chrome as Legitimacy, Not Excess

For most of Cadillac's history, chrome was not decoration. It was a declaration.

Chrome signaled durability before a word was spoken. It reflected light not to attract attention, but to announce presence. It said this object was meant to last, meant to be maintained, meant to occupy public space with confidence. In American luxury, chrome was not a flourish. It was a promise.

This distinction matters because modern eyes often misread chrome through a contemporary lens—associating it with excess, nostalgia, or ornamentation. That reading ignores the cultural role chrome played when Cadillac established its authority.

In the early 1990s, chrome and Cadillac were still natural allies. The visual language of American luxury was rooted in reassurance, not performance signaling. Authority was meant to be seen from a distance and understood instantly. A Cadillac did not need to explain itself. It needed to be recognized.

Chrome did that work.

It conveyed permanence in an era when institutions were trusted. It conveyed industrial confidence in a culture that believed manufacturing signaled seriousness. It conveyed prosperity that was visible, earned, and shared. Postwar America did not hide success. It displayed it calmly.

This is why chrome functioned as reassurance rather than flash. It did not shout. It asserted. It told owners and observers alike that the car was substantial, established, and not provisional. The vehicle was not trying to impress. It was confirming its place.

That difference—between assertion and ornamentation—is where many modern critiques miss the point.

Chrome as ornament is decoration layered onto an object. Chrome as assertion is part of the object's identity. One is optional. The other is load-bearing.

Cadillac's use of chrome in this period was structural. It framed the car. It marked edges and boundaries. It highlighted scale and proportion. It did not compete with the form; it clarified it. The chrome said, "This is where the car begins and ends. This is what it weighs. This is how seriously it should be taken."

This is why something resembling an AMG Hammer[1], visually transplanted into early-1990s Cadillac culture, would have been a nonstarter for most owners. Not because they rejected performance, but because the visual language would have broken trust. Monochrome severity would have read as absence, not discipline. The car would have looked unfinished, not focused.

Performance without reassurance was culturally incoherent.

Cadillac's later shift toward sport-oriented visual language—reduced chrome, darker finishes, sharper edges—did not emerge in a vacuum. It followed a broader cultural change. Institutional trust eroded. Prosperity became more private.

---

[1]https://www.hagerty.com/media/car-profiles/amg-hammer-wagon/

Performance began to replace presence as a marker of legitimacy. By the time Northstar-powered STS models appeared, the market was ready to read restraint as intent rather than omission.

But that was not yet true in the early 1990s.

At that moment, Cadillac authority still depended on visible affirmation. The customer definition of luxury included calm, space, and certainty. Chrome communicated all three. Removing it did not make the car more serious. It made it less legible.

This is why the Allanté is such a critical reference point.

The Allanté represented restraint without denial. Designed by Pininfarina, it spoke a quieter design language, but it did not abandon Cadillac's visual grammar. Chrome remained present—controlled, proportional, intentional. It signaled family membership. The car could be modern and international without severing its lineage.

That is exactly the kind of precedent an AMG-involved Cadillac project would have studied.

AMG would not have imposed monochrome as a matter of taste. They would have asked what restraint meant within Cadillac's visual culture. The answer would not have been removal. It would have been refinement. Chrome reduced, not erased. Brightwork disciplined, not denied. Authority preserved, not hidden.

This is how legitimacy is maintained during evolution.

Chrome, in this context, is not about nostalgia. It is about continuity. It is a visible link between past authority and future capability. It reassures the audience that while the system beneath the surface may have changed dramatically, the promise has not been broken.

In systems terms, this is backward compatibility made visible.

Legitimacy depends on recognizability. When users cannot recognize authority, they do not grant it patience. They challenge it early. They interrogate it constantly. They resist it reflexively. Chrome reduced that friction in American luxury by making authority obvious.

The lesson is not that chrome is always correct. It is that legitimacy must be expressed in a language the audience already trusts.

In American luxury, authority was meant to be seen. Not flaunted. Not hidden. Seen.

Cadillac earned its place by making permanence visible. Any attempt to modernize that authority without respecting its visual foundations risks breaking the interface that made the authority legible in the first place.

This is why chrome was not excess. It was infrastructure.

And infrastructure, once removed without replacement, reveals how much load it was carrying all along.

Monochrome as Discipline in a German Context

On the Autobahn, authority announces itself differently.

There is no horn blast. No flourish. No chrome glinting in the sun. There is only a shape in the mirror—dark, compact, accelerating with calm intent. A monochromatic dot that grows quickly, steadily, without drama. You move over not because you are intimidated, but because you understand what you are seeing.

That is monochrome as discipline.

In the German performance context, monochrome is not an aesthetic choice. It is a declaration of internal confidence. It signals that the system does not require reassurance, explanation, or validation. The car is not asking to be recognized. It is assuming recognition.

AMG's visual language emerged from this assumption.

AMG vehicles, especially in their formative years, did not present themselves as luxury objects in the American sense. They presented themselves as serious machines operated by serious people. The subdued palette, reduced brightwork, and restrained badging were not meant to hide capability. They were meant to refuse performance as spectacle.

Monochrome said: *If you know, you know.*

This worked because the surrounding culture supported it. German automotive authority is not negotiated at first contact. It is presumed. The Autobahn itself encodes this assumption. High-speed travel is normalized. Competence is expected. Restraint is read as professionalism, not absence.

In that context, understatement is not humility. It is confidence so internalized that it no longer needs reinforcement.

Monochrome communicates refusal on multiple levels. It refuses ornament. It refuses distraction. It refuses to perform for anyone who is not already fluent in the language of the machine. This refusal is not hostile. It is clarifying. It draws a boundary around the intended audience.

The message is simple and complete:

> "We know what this is. We are not explaining ourselves."

That message lands because authority is already established upstream—by engineering standards, by institutional trust, by shared expectations of competence. Visual restraint becomes legible as discipline precisely because the audience does not require reassurance.

This is why monochrome works where authority is assumed, not negotiated.

AMG could strip visual signals without weakening legitimacy because legitimacy was already present. The brand did not need to prove seriousness through size, shine, or spectacle. It proved seriousness through behavior: composure at speed, stability under load, repeatable performance without drama.

The visual language followed the behavior. It did not lead it.

This relationship is critical. Monochrome does not create authority. It reflects authority that already exists. Applied prematurely, it does not read as discipline. It reads as vacancy.

This is where translation fails.

When monochrome is applied in a culture where authority must first be demonstrated visually, the signal collapses. What reads as restraint in one context reads as omission in another. What reads as seriousness reads as cost-cutting. What reads as internal confidence reads as indifference.

The signal does not carry because the assumptions do not match.

American luxury historically did not assume authority. It performed it. Not theatrically, but visibly. Presence mattered because the audience expected authority to be announced before it was exercised. Size, brightness, and finish did not undermine seriousness. They established it.

German performance culture evolved differently. Authority was embedded in systems, not surfaces. The visual language could afford to retreat because trust was already institutionalized.

This difference is not a matter of taste. It is a matter of interface.

Interfaces work only when sender and receiver share context. Monochrome is a dense signal. It compresses meaning rather than expanding it. That compression only works when the receiver knows how to decompress it.

On the Autobahn, they do.

In early-1990s American luxury, most did not.

This is why an AMG-style monochrome treatment applied wholesale to a Cadillac Fleetwood would have failed—not mechanically, but semantically. The car would have spoken a language its audience did not yet understand. Authority would not have been recognized, even if it were present.

AMG would have understood this. Their discipline was never about imposing a look. It was about aligning signals with reality. In contexts where authority was negotiated visually, they would not have removed the visual cues that made that negotiation possible.

Monochrome was never the point. Discipline was.

The mistake modern systems make is copying the surface without inheriting the context. They adopt minimalism, restraint, and opacity believing these qualities signal seriousness. But without shared assumptions, those signals do not compress meaning—they erase it.

Monochrome, without authority, is silence.

This is the core lesson of AMG's visual discipline. It is not a style to be transplanted. It is a language to be spoken only where it will be understood. Where authority is assumed, monochrome sharpens it. Where authority must be earned, monochrome delays it.

Nothing says "move over" on the Autobahn like a fast-growing monochromatic dot in the mirror. But that signal works because everyone involved already agrees on what it means.

Authority, in that context, does not ask. It proceeds.

And that is why monochrome works—precisely where authority does not need to explain itself.

When Correctness Reads as Absence

The most dangerous failure mode for authority is not being wrong. It is being right—and unreadable.

Systems rarely fail because they violate their own rules. They fail because they speak in a language their audience cannot interpret. The engineering is sound. The discipline is real. The intent is coherent. And yet the system is quietly rejected, not through revolt but through unease. Something feels missing, even when nothing is technically absent.

This is where correctness reads as absence.

In the context of our thought experiment, this is the moment an AMG-influenced Fleetwood risks losing its audience—not because it fails to perform, but because it fails to announce its performance in a culturally legible way. The structure may be disciplined beneath the surface, but the surface no longer communicates authority in the way Cadillac buyers of the early 1990s were trained to recognize.

A monochrome Fleetwood, stripped of brightwork and visible reassurance, would not read as focused. It would read as unfinished. The cues that historically told owners "this car is substantial, established, and intentional" would be gone, replaced by restraint that had not yet earned trust in that cultural context.

The discipline would be invisible. And invisible discipline is indistinguishable from neglect.

This is not an argument against correctness. It is an argument against assuming correctness is self-evident.

AMG's monochrome language worked because it operated inside an ecosystem where authority was already institutionalized. German performance culture did not require reassurance through surface cues because seriousness was presumed. When signals were removed, the audience understood the absence as a choice, not a deficiency.

Cadillac's audience, at that moment in time, did not share those assumptions.

American luxury had been built on visible affirmation. Presence mattered. Scale mattered. Brightness mattered—not as ostentation, but as confirmation. Authority was meant to be recognized immediately, not inferred after analysis. Removing those signals prematurely broke the interface between the system and its audience.

The same failure pattern appears relentlessly in modern software systems.

Technically correct platforms fail adoption not because they are flawed, but because they are illegible to the people expected to trust them. Interfaces optimized for internal elegance read as indifference externally. Minimal dashboards feel evasive.

Sparse reporting is interpreted not as curation, but as withholding.

The system is correct. The interface is wrong.

Executives encountering such systems often struggle to articulate their discomfort. There is no obvious defect to point to. No missing feature. No clear error. And yet confidence erodes. Decisions slow. Workarounds appear. Escalation replaces trust.

Correctness without legibility does not inspire confidence. It provokes interrogation.

This dynamic is not new.

In the 1980s and 1990s, a distinct role existed to prevent exactly this outcome: the systems analyst. That role acted as a bridge between business intent and technical execution. The systems analyst worked with stakeholders to understand what actually mattered—the decision to be supported, the risk to be managed, the outcome that carried value—and then worked backward to define the minimal viable structure required to deliver that understanding.

The emphasis was not on building everything that could be built. It was on delivering what needed to be recognized.

The systems analyst translated meaning. They ensured that what was delivered would be legible to the audience expected to rely on it. Authority was preserved not through completeness, but through alignment.

As development practices evolved, that role quietly dissolved.

Agile brought speed, adaptability, and local optimization. It excelled at producing increments of functionality. But in many organizations, it replaced deep audience understanding with emergent design. Meaning was allowed to surface gradually, inferred from iterations rather than declared up front.

The result is familiar.

Systems filled with graphs, charts, and metrics that are almost what the business was looking for—but not quite. They are data-rich. They are meticulously instrumented. They are technically impressive. And yet they leave decision-makers uneasy.

Douglas Adams described something as "almost, but not quite, entirely unlike" the thing it was meant to be. Modern software often lands in that same space. It gestures toward insight without quite delivering it. It looks authoritative without quite being legible.

The discomfort this creates mirrors another phenomenon: the uncanny valley. High-grade animation that is almost human, but not quite, provokes unease rather than empathy. The closer it gets without fully crossing the threshold, the more unsettling it becomes.

Delivered software can fall into the same valley.

It is realistic enough to inspire expectation, but misaligned enough to betray it. Authority feels implied but not earned. Correctness is present, but legitimacy is not.

This is the tragedy of un-translated discipline.

In automotive terms, it is AMG-level rigor applied without Cadillac's visual grammar. The work is sound. The intent is good. But the audience does not know how to read it. Something feels missing, even if nothing has been removed mechanically.

Cadillac understood that authority had to be visible to be granted. AMG understood that restraint only works when authority is already assumed. Both understood the same truth: discipline must be expressed in a language the audience can decode.

Systems that ignore this reality often believe they are being serious when they are actually being opaque. They mistake silence for confidence. They assume that correctness will eventually explain itself.

It rarely does.

Authority that must be decoded will be resisted. Authority that is recognized will be granted room to operate. When correctness reads as absence, legitimacy fails first—and performance follows shortly thereafter.

This is why appearance, whether in cars or systems, is never cosmetic. It is how discipline introduces itself to the world. It is the interface through which authority becomes legible.

And authority that cannot be recognized is authority that will not be trusted, no matter how correct it may be underneath.

Legitimacy Is Load-Bearing

Legitimacy is often treated as a feeling—something soft, psychological, or political. In practice, it is structural. It carries load.

Systems that are perceived as legitimate move through the world with less resistance. Decisions pass more easily. Exceptions are granted more readily. Ambiguity is tolerated longer. This is not because the system is indulged, but because it is trusted to be acting within understood bounds.

That trust does real work.

Legitimacy reduces friction before interaction. It absorbs doubt before it hardens into challenge. It prevents interrogation from becoming the default posture. A legitimate system is not immune from scrutiny, but it is not required to re-earn the right to operate at every step.

This is why legitimacy must be designed, not hoped for.

In the automotive world, legitimacy is immediately tangible. A flagship sedan that looks authoritative is given space—literally and figuratively. Other drivers yield. Owners assume competence. Minor flaws are contextualized rather than amplified. The vehicle is granted patience because it appears to belong in its role.

That patience is load-bearing.

Without it, every interaction becomes adversarial. The car is questioned before it is experienced. Its behavior is interpreted skeptically. Its deviations are magnified. The same mechanical performance now feels fragile because the surrounding trust has collapsed.

Cadillac understood this intuitively for decades. Its visual language reduced friction long before any mechanical claim was tested. Presence, proportion, and finish communicated that the car was meant to carry responsibility—executive, institutional, familial. That message mattered because it framed how everything else would be judged.

Legitimacy did not make the car better. It made the car believable.

The same dynamic governs systems.

In software and infrastructure, legitimacy reduces verification cost. Teams trust systems that look familiar, intentional, and aligned with their expectations. They do not need to constantly cross-check, second-guess, or demand additional proof. The system's outputs are accepted as provisional truth unless contradicted.

This dramatically lowers cognitive load.

Familiar signals—terminology, structure, visual hierarchy—allow users to orient themselves quickly. They know where to look, what matters, and which signals are authoritative. When those cues are present, attention is conserved for decisions rather than spent on interpretation.

When they are absent, everything becomes harder.

Illegitimate systems force users to translate constantly. Every graph must be decoded. Every alert must be contextualized. Every report invites a follow-up question. The system may be correct, but it is exhausting. Over time, exhaustion turns into escalation.

Culturally aligned interfaces prevent this escalation by making intent legible. They tell the user, "This system understands your role. It knows what authority you have and what you need to see." That alignment is not cosmetic. It is structural. It determines how much stress the system can absorb before it fractures under scrutiny.

Legitimacy is what allows systems to survive pressure without reconfiguration.

This is why legitimacy precedes governance. Rules are easier to enforce when the system enforcing them is trusted. Constraints are accepted more readily when they feel intentional rather than arbitrary. Refusal is tolerated when it comes from a system that appears grounded and serious.

Without legitimacy, even correct refusal feels hostile.

This is where many modern platforms falter. They focus on correctness, scalability, and efficiency while neglecting legibility. They assume legitimacy will emerge from performance alone. When it does not, they respond by adding explanation—more dashboards, more documentation, more metrics—rather than addressing the interface itself.

Explanation is not a substitute for legitimacy.

Legitimate systems do not need to explain themselves constantly. They communicate their intent through structure. Users understand what they are dealing with before interaction, and that understanding shapes how feedback is interpreted.

This is load-bearing because it determines how much ambiguity a system can tolerate. All systems encounter moments where data is incomplete, signals conflict, or outcomes are delayed. Legitimate systems are given room to resolve these moments. Illegitimate systems are escalated immediately.

The difference is not technical. It is perceptual—and perception is structural.

Returning to our Fleetwood, this is why visual legitimacy could not be discarded in pursuit of discipline. Removing chrome without replacing its signaling function would have increased friction everywhere else. Owners would question decisions sooner. Deviations would feel like failures. The car would be asked to justify itself constantly.

That burden would not show up in a specification sheet. It would show up in trust.

AMG would have understood that legitimacy had to be preserved even as discipline increased. The visual language would need to carry authority forward so the underlying changes could do their work without being challenged at every turn.

The same principle applies to systems undergoing modernization. When interfaces change faster than trust can adapt, legitimacy collapses. Users respond by imposing their own controls—shadow dashboards, manual checks, parallel reporting. The system becomes heavier, not lighter.

Legitimacy, when present, prevents this accretion of defensive behavior.

This is why legitimacy is not optional. It is not branding. It is not marketing. It is infrastructure. It carries load that would otherwise be borne by people, processes, and politics.

When legitimacy is designed correctly, systems move with less resistance. When it is neglected, no amount of correctness will compensate.

Legitimacy is not cosmetic. It carries load.

Not all machine-driven optimization represents a loss of authority.

There is an important distinction between systems that optimize in response to load and systems that explore configuration space under declared governance.

Self-optimizing platforms such as container orchestration systems primarily react to pressure. They scale, reschedule, and rebalance in pursuit of continued operation. Their objective function is survival. Load becomes the governing signal, and authority follows demand.

By contrast, systems such as Akamas operate under an explicitly declared objective. Humans define the outcome—latency, throughput, stability, cost, or error rate—and the machine explores the configuration space within those constraints. The goal does not move. The policy envelope remains fixed.

This distinction matters because the configuration space of modern platforms is no longer human-navigable. A contemporary Java runtime exposes hundreds of interdependent parameters. The total combination space exceeds what any team can explore on a human timescale. In such environments, brute-force cycling combined with machine learning is not abdication—it is the only practical method of discovery.

Authority is preserved because optimization is bounded. The machine is permitted to search, not to redefine success. It accelerates exploration without assuming jurisdiction.

Optimization becomes dangerous only when the system is allowed to move its own goals. Exploration under policy is governance. Adaptation without policy is surrender.

Making the Fleetwood an American Flagship Again

By this point in the book, the Fleetwood is no longer a curiosity. It is a fully specified thought experiment. The frame has been made authoritative. The brakes refuse dishonesty. The suspension mediates truth. Time has been disciplined. Governance has been separated from feedback. Legitimacy has been restored as a structural concern.

Now the question becomes simple, and unavoidable:

What does this car become?

The answer is not a German transplant wearing American sheet metal. And it is not nostalgia, frozen and polished. It is something rarer and harder to execute: a credible American executive flagship, rebuilt with discipline but fluent in its own culture.

To do that, some things must remain non-negotiable.

What stays Cadillac is not a parts list. It is a posture.

Presence remains. The car must still occupy space calmly and without apology. It should not look eager. It should not look aggressive. It should look inevitable. Long hood, long deck, measured proportions. A vehicle that signals responsibility before it signals motion.

Calm remains. Not softness, but composure. The ride does not communicate everything the road is doing. It communicates what matters. The driver is informed without being burdened. The cabin is a place of decision, not stimulation.

Visual authority remains. The car must still announce itself as something meant to carry weight—organizational, familial, institutional. Chrome does not disappear. It is refined. Disciplined. Purposeful. It continues to mark edges, scale, and seriousness. Authority is still meant to be seen.

These are not indulgences. They are the interface.

What AMG refines lives beneath that interface.

Discipline replaces tolerance. The frame no longer yields quietly. It holds geometry under load. Body and structure work together deliberately, not accidentally. Compliance exists where it is chosen, not where it happens to remain.

Composure replaces isolation. The car does not float away from reality; it translates it. Steering weight returns meaning. Braking becomes repeatable and authoritative. Motion is controlled rather than absorbed.

Restraint beneath the surface becomes the defining characteristic. Power is present, but never theatrical. Torque arrives predictably. The drivetrain does not announce itself. It simply does its work. Performance is something the car possesses, not something it performs.

This is where the AMG influence is most honest. Not in visual mimicry, but in behavioral integrity. The car does not pretend to be something it is not. It simply stops lying to itself.

The result is neither German nor nostalgic.

It is not a monochrome statement of internal confidence that assumes institutional authority. That would misread the culture. And it is not a throwback to an era when size and softness alone defined luxury. That would ignore what has been learned.

Instead, it becomes an executive express in the American sense: a vehicle designed to move people who carry responsibility, with calm, certainty, and dignity. A car that does not ask to be admired, but expects to be trusted.

Importantly, this Fleetwood does not need to explain itself. Its legitimacy is legible. Its restraint is expressed, not hidden. Its authority is recognized before it is tested.

That recognition matters because it determines how everything else is received. Deviations are contextualized. Decisions are trusted. The system is granted patience.

This is the same dynamic that governs successful systems.

When modernization respects cultural interfaces, authority survives change. When discipline is translated rather than imposed, trust is preserved. When legitimacy is treated as load-bearing, performance can improve without triggering resistance.

The failure mode you have been circling throughout this chapter is not technical. It is translational. It is the belief that correctness alone will be enough, that discipline can remain invisible, that authority will announce itself eventually.

It will not.

Authority that cannot be recognized will not be granted. Authority that speaks its culture will not need to ask.

That is the lesson of this Fleetwood that was never built—and the systems that fail because they forget the same truth.

Legibility is a form of authority.

7

When Authority Cannot Emerge

The Myth of Emergent Authority

Complex systems are often mistaken for self-governing ones.

They produce patterns. They stabilize. They converge on behaviors that appear consistent, repeatable, and even successful. From the outside, this looks like authority taking shape—decisions aligning, outcomes reinforcing themselves, disagreement smoothing out over time.

This is an illusion.

Authority does not emerge naturally from complexity. What emerges are equilibria—local balances of force, incentive, habit, and tolerance. These balances can persist for long periods without ever becoming truthful, just, or safe.

A system can be orderly and still be ungoverned.

Stability is not proof of authority. It is merely proof that the system has found a way to continue.

This distinction matters because many organizations—technical, operational, and institutional—rely on emergence as a substitute for decision. They assume that if enough competent people interact long enough, authority will surface on its own. That disagreement will resolve itself. That correctness will assert gravity over time.

It rarely does.

What emerges instead are patterns optimized for survival within the system as it exists—not for truth, not for correctness, and not for responsibility.

Agreement is often mistaken for authority because it feels legitimate. When many people align on an outcome, it appears justified. But agreement only reflects convergence, not mandate. It tells us that perspectives have been smoothed, not that responsibility has been assumed.

Consensus is a social artifact. Authority is a structural one.

A group can agree completely and still have no one accountable for the consequences of that agreement. In such cases, authority has not been established—it has been diffused. Responsibility has been spread thin enough that it disappears.

The same confusion applies to coherence.

Systems often develop internally consistent narratives about why they behave the way they do. Processes align. Metrics reinforce one another. Explanations fit neatly together. This coherence is comforting. It creates the appearance of correctness.

But coherence is not truth.

A system can be perfectly coherent while being fundamentally wrong. Internally consistent assumptions can reinforce one another indefinitely, especially when no external authority exists to challenge them. In such environments, dissent feels disruptive not because it is incorrect, but because it threatens stability.

Order becomes self-protective.

This is how ungoverned systems persist. They do not collapse immediately. They normalize. They adapt around flaws. They absorb contradictions by redefining success. Over time, the absence of authority becomes invisible, replaced by ritual, habit, and precedent.

Emergence excels at producing this kind of order.

But emergence does not produce responsibility.

Responsibility requires a boundary—someone who can say this stops here, this changes now, this is no longer acceptable. Emergent systems resist such boundaries because boundaries interrupt pattern formation. They force choices that cannot be smoothed over by consensus or delayed by interpretation.

This is why authority cannot arise naturally in complex systems.

Authority requires declaration.

It requires someone—or some explicitly defined body—to accept obligation for outcomes, not just participation in process. It requires the ability to be wrong publicly, decisively, and without deflection. These properties do not emerge. They must be imposed.

The Fleetwood program made this refusal explicit.

The vehicle was not allowed to discover its identity through iteration, testing, or compromise. There was no exploratory phase in which character could "emerge" from tuning, market reaction, or accumulated performance data. Before a single component was chosen, the end state was declared.

Silence under load. Effortless torque. Sustained authority without agitation.

These were not goals to be validated later. They were constraints imposed at the start.

In this sense, the program did not wait for authority to surface. It imposed authority deliberately, before the system could begin negotiating with itself.

Systems without declared authority often appear functional for long periods. They ship. They scale. They meet targets. This apparent success reinforces the belief that authority is unnecessary—that performance itself has validated the structure.

This belief is dangerous.

Performance only tells us that the system has not yet encountered conditions that exceed its tolerance. It says nothing about how the system will behave when those conditions arrive. Without authority, there is no mechanism to decide what must change when success itself becomes the problem.

Emergent systems are particularly vulnerable here. Because no one owns the decision, the system defaults to continuation. Past success becomes justification for future inaction. Drift accelerates quietly, shielded by metrics that still look acceptable.

The system does not fail loudly.

It fails by becoming unable to choose.

This is the central failure mode of emergent authority: not chaos, but sanctioned ambiguity. Everyone participates. No one decides. The system moves forward because movement is easier than refusal.

Order persists. Authority does not.

It is important to be precise here. This is not an argument against collaboration, learning, or adaptation. Emergence is powerful at generating insight, surfacing options, and exploring solution space. But it cannot close that space. It cannot declare truth. It cannot accept responsibility.

Those acts are discontinuous. They require interruption.

Authority begins where emergence must end.

The belief that authority will eventually "settle" if we wait long enough—if we gather more data, run more experiments, include more voices—is not humility. It is abdication. It shifts responsibility from decision to process and mistakes motion for governance.

Delay feels safe because it avoids error.

In reality, delay selects outcomes by default. Existing power structures harden. Existing assumptions calcify. Existing risks compound silently. The absence of authority does not suspend decision—it hands it to inertia.

Non-decision is not neutrality.

It is an active governance posture with consequences.

This is why authority must be declared explicitly in systems that matter. Not because certainty has been achieved, but because it has not. Not because disagreement has ended, but because it persists. Not because the system has failed, but because waiting for failure is itself a choice.

Emergence produces patterns.

Authority produces responsibility.

Confusing the two is how systems remain orderly while drifting into failure.

Why Substitutes for Authority Cannot Decide

In complex systems, authority is uncomfortable.

It carries consequence. It exposes judgment. It requires someone to say *this will stand*—not because it is provably optimal, but because responsibility must land somewhere. When that burden feels risky, systems reach for substitutes. They look for mechanisms that feel safer, more neutral, more defensible.

Data. Consensus. Performance. Time.

Each is mistaken, repeatedly, for authority. Each substitute fails for a different reason.

Each feels earned. Each feels objective. Each feels legitimate.

None of them can decide.

This confusion is not accidental. It is structural. Modern systems are extraordinarily good at measuring, aligning, producing, and persisting. They are far less comfortable declaring obligation under uncertainty. The result is a culture that *appears* rigorous while quietly avoiding the act that governance requires: A decision.

Data as a Surrogate for Authority

Data is often invited to replace judgment.

When systems become complex enough that responsibility feels dangerous, measurement multiplies. Dashboards expand. Metrics become more granular. Confidence intervals tighten. The hope is implicit and persistent: if we observe enough, the correct decision will emerge on its own.

It does not.

Data describes behavior. It does not select outcomes.

This distinction is blurred because data feels objective. Numbers appear neutral. Charts look factual. But measurement does not collapse uncertainty—it expands it. Each new metric introduces another interpretive axis. Each additional data source multiplies explanations rather than resolving them.

More data rarely ends disagreement. It professionalizes it.

This is not a failure of analytics. It is a category error.

Data can tell us *what is happening*. It cannot tell us *what must be done*.

Selection is not an analytical act. It is a declarative one.

Metrics narrow attention. They reveal gradients, trends, and deviations. They illuminate terrain. But they do not grant permission. They do not authorize action. They do not accept consequence. They cannot be wrong in a way that matters because they cannot be held accountable.

This becomes dangerous when waiting for more data is mistaken for prudence. The language shifts subtly: *we need more signal, the sample size isn't large enough, let's wait for another cycle.* Each phrase sounds reasonable. Together, they form a posture.

Delay dressed as rigor.

By the time data is unambiguous, the system has usually already decided. Architecture has hardened. Costs are embedded. Behavior is normalized. At that point, measurement does not enable decision—it explains why decision is now expensive.

Data does not fail by misleading. It fails by being silent on obligation.

Consensus as a Substitute for Legitimacy

It is postponed.

Governance reference: Authority does not emerge. It is declared, scoped, and enforced. See Chapters 1–2.

Performance as Retroactive Permission

Performance is often mistaken for earned authority.

When systems ship, scale, or succeed, outcomes begin to carry weight beyond their proper role. Success is treated not merely as evidence of competent execution, but as retroactive justification for the right to decide.

This belief is convenient.

It allows authority to be inferred rather than declared. It replaces responsibility with momentum. It postpones governance while results remain acceptable.

But performance does not grant authority.

Performance validates outcomes, not decisions.

Every system performs according to structure already embedded within it. Architecture precedes throughput. Boundaries precede behavior. Incentives precede output. By the time performance can be measured, authority—or its absence—has already done its work.

Success is a lagging signal.

A system can perform well while accumulating hidden risk. In fact, success is often the most fertile environment for risk accumulation. When outcomes are positive, scrutiny softens. Temporary exceptions harden into structure. Constraints drift quietly out of view.

Performance creates cover.

Warnings raised during periods of success are reframed as theoretical. Concerns about scalability, security, or long-term cost are postponed in favor of maintaining momentum. The system appears healthy precisely because it has not yet been asked to endure conditions that exceed its tolerance.

Performance delays scrutiny. It does not resolve it.

This is why high-performing systems often fail abruptly rather than gradually. The signals were present, but success suppressed their urgency. By the time performance degrades, the architecture is fixed and options are narrow.

Performance did not prevent failure. It concealed it.

Treating success as jurisdiction produces systems that are confident without being governed, productive without being safe, and successful right up until they are not.

Evidence informs authority. It does not create it.

Time as Accumulated Entitlement

Time is often mistaken for mandate.

In long-lived systems, tenure slowly accrues weight. Those who have been present longest are assumed to have earned stronger decision rights. Familiarity with history is conflated with jurisdiction. Survival is treated as evidence of correctness.

It is not.

Longevity creates familiarity, not authority.

Time teaches adaptation. It teaches how to navigate incentives, personalities, and constraints. It teaches which questions are rewarded and which ones carry cost. Over time, participants become fluent in how the system behaves.

This fluency is valuable. It is not governance.

Experience increases confidence far more reliably than it increases correctness. As systems drift, long-tenured participants drift with them. Exceptions normalize.

Temporary workarounds become invisible. What once required justification becomes background noise.

Time rewards adaptation, not challenge.

This creates a dangerous inversion: those most trusted to decide are often those most acclimated to accumulated compromise. Their intuition is shaped by precedent rather than obligation.

From inside the system, everything feels reasonable. From outside, deviation is obvious.

This is why time cannot decide.

A system can persist for years without being governed. It can drift steadily away from its original intent while appearing stable to those within it. Tenure becomes a veto not because it is correct, but because it is old.

Authority is not memory. Authority is obligation.

Remembering why something evolved does not justify its continuation. Knowing where the scars are buried does not grant permission to add new ones.

Experience informs decisions. It does not own them.

Authority Cannot Be Outsourced

Data, consensus, performance, and time all feel legitimate because they reduce personal risk. They diffuse responsibility. They allow decisions to appear inevitable rather than chosen.

That is precisely why they fail.

None of these mechanisms can accept consequence. None can be wrong in a way that matters. None can answer for outcomes.

Authority can.

Authority is inseparable from accountability. It exists only where someone accepts that they may be wrong—and will still stand behind the decision. It cannot be delayed until uncertainty disappears, because uncertainty is the condition in which decisions exist.

Waiting for substitutes to decide is not rigor. It is refusal.

Systems do not become safe when uncertainty is eliminated. They become safe when uncertainty is acknowledged and governed.

Authority must be declared explicitly and early—before data accumulates, before consensus forms, before performance succeeds, before time normalizes drift. Only then can these mechanisms serve their proper role.

Data informs authority. Consensus follows authority. Performance validates authority. Experience constrains authority.

None of them replace it.

When authority is absent, the system will still decide. It will do so through inertia, drift, incentives, and accumulated habit. And it will do so without obligation, memory, or restraint.

That is not governance.

That is abdication made orderly.

Authority is not comfortable. That is how you know it is real.

Delay as an Active Decision

Delay is often mistaken for caution.

When authority cannot be comfortably declared, systems frequently choose to wait. The language sounds responsible: *we need more information, the timing isn't right, let's see how this develops*. Waiting is framed as neutrality—a temporary suspension of commitment until uncertainty resolves itself.

This framing is false.

Delay is not the absence of decision. Delay is a decision to allow other forces to govern.

Non-decision is not neutrality. It is delegation to entropy.

Every moment without declared authority selects outcomes by default. Existing structures continue operating. Existing incentives remain unchallenged. Existing assumptions harden quietly. What is already in motion gains advantage simply by persisting.

Inaction privileges incumbency.

This is not an abstract claim. It is a mechanical one. Systems without intervention follow the path of least resistance, not the path of correctness. Delay transfers authority from intent to inertia.

Waiting is therefore a governance posture.

It determines who benefits, which options remain viable, and which risks are allowed to compound. The longer authority is withheld, the fewer choices remain reversible. By the time action is finally taken, the system has often already decided.

This is the danger of delay disguised as prudence.

In technical systems, this failure mode is common. Teams wait for certainty that cannot arrive. They postpone hard declarations until the cost of change becomes prohibitive. What began as restraint becomes irreversibility. The system appears calm while quietly closing doors.

The Fleetwood program confronted this directly at its most sensitive point: the engine.

From the outset, the engine was acknowledged as the central unknown. No existing GM or Cadillac powerplant satisfied the declared constraints of character, mass, balance, and torque delivery. This was not disputed. What was at risk was how that uncertainty would be governed.

The easy path would have been delay without containment.

Waiting for a future engine revision. Waiting for another internal option to mature. Waiting for consensus. Waiting for proof.

Each of those options would have allowed the system to drift toward convenience. Existing platforms would have gained weight simply by being available. The vehicle's identity would have been negotiated indirectly through packaging constraints, weight distribution compromises, and late-stage accommodations.

Nothing would have been "decided."

Everything would have been selected by default.

That outcome was explicitly rejected.

The engine was left unresolved, but it was not left ungoverned.

Its uncertainty was named. Its scope was bounded. Its jurisdiction was isolated. The decision to defer was itself declared and structured. Delay was not permitted to operate invisibly.

This distinction matters.

There is a difference between *postponing a choice* and *allowing the system to choose for you*. The first is disciplined. The second is abdication.

By treating delay as an active decision, the Fleetwood program preserved authority even in the absence of resolution. The engine question did not leak into unrelated domains. It did not justify compromises elsewhere. It did not accumulate silent power simply because time was passing.

Delay was contained.

This is rarely how delay operates in software systems.

More commonly, unresolved decisions are allowed to bleed outward. Architectural ambiguity becomes permanent. Temporary scaffolding hardens. Early assumptions gain authority through survival rather than validation. By the time a decision is forced, the system can no longer tolerate it.

The system waited itself into constraint.

In such cases, the cost of delay is not immediately visible. It appears later as brittleness, expense, or catastrophic inflexibility. Teams are surprised by how few options remain, forgetting that those options were surrendered incrementally.

Delay did not preserve optionality.

It consumed it.

This is why delay must be treated as an explicit act of governance. If authority cannot yet be declared, the correct response is not passive waiting, but active containment. Boundaries must be enforced. Scope must be frozen. Drift must be resisted.

Otherwise, the system will continue evolving under forces that have no obligation to truth, safety, or intent.

Every moment without authority is a moment governed by something else.

In the Fleetwood program, this recognition marked the end of abstract discussion and the beginning of disciplined construction. Authority had now been asserted over what could decide—and what could not. Delay was no longer invisible.

Only after that condition was met could building responsibly begin.

The Obligation to Decide

Authority does not require certainty.

This is the final misconception that must be discarded before construction can begin.

Many systems delay decision because they believe authority is only legitimate when confidence is complete—when uncertainty has been reduced to a tolerable minimum, when evidence feels sufficient, when disagreement has quieted. In this view, decision is treated as a reward earned through analysis, consensus, performance, or time.

This view is wrong.

Authority exists precisely because certainty does not.

If certainty were available, authority would be unnecessary. Decisions would collapse into deduction. Responsibility would evaporate into inevitability. The presence of authority is a signal that the system has reached a boundary where evidence ends and obligation begins.

Decision is not the reward for being right.

Decision is the acceptance of being answerable if wrong.

This is why refusal to decide is not restraint. It is abdication.

When authority holders decline to decide, authority does not pause. It dissolves. Responsibility migrates to forces that cannot be held accountable: inertia, convenience, incumbency, momentum. The system continues moving, but no longer under governance.

At that point, authority has already collapsed—quietly, without ceremony.

The most damaging failures in complex systems do not occur because someone decided incorrectly. They occur because no one accepted the obligation to decide at all. In those environments, outcomes are produced, but no one owns them. Harm accumulates without attribution. Drift becomes indistinguishable from intent.

The system functions. Authority does not.

This is why decision is not optional.

Authority is not something that emerges when the moment feels safe. It is something that must be asserted precisely when the moment is unsafe—when information is incomplete, when consequences are unclear, when disagreement persists.

Waiting for certainty does not preserve authority.

It guarantees its collapse.

In the Fleetwood program, this obligation was recognized explicitly. The absence of a resolved engine did not suspend authority. It sharpened it. The decision to proceed with construction boundaries intact—to isolate the unknown rather than allow it to govern—was not made because confidence was high.

It was made because responsibility could not be deferred.

Authority was exercised not to eliminate uncertainty, but to contain it.

This is the distinction that separates governance from optimism.

Decision is not an act of confidence. Decision is an act of custody.

Someone must accept stewardship of the outcome before the system can be trusted to move forward. That acceptance cannot be crowdsourced. It cannot be automated. It cannot be postponed until conditions are comfortable.

Authority does not wait to be earned.

It must be assumed.

This does not mean every decision is final. It does not mean authority is infallible. It does not mean dissent disappears. It means that someone stands where consequence will land.

Without that, systems do not fail loudly.

They fail by default.

Someone must decide—even without certainty—or authority collapses by default.

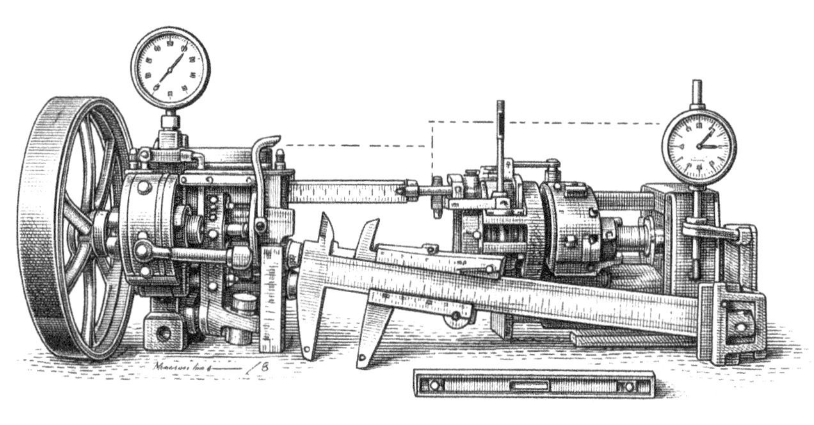

Authority cannot emerge from incoherence.

8

Jurisdiction is not a side effect

The Four Roles That Are Always Confused

Before authority can be discussed, four roles must be separated.

They are often spoken of interchangeably. In practice, they are treated as interchangeable. In functioning systems, they frequently coexist within the same individuals. Structurally, however, they are never the same thing.

Confusing them is how systems become busy without being governed.

The first is **responsibility**.

Responsibility describes who performs the work. It is operational. It answers the question: *who does the thing?* Engineers design components. Craftsmen build interiors. Suppliers deliver assemblies. Responsibility is visible, measurable, and typically well understood. It is where effort lives.

The second is **accountability**.

Accountability describes who answers for outcomes. It is retrospective. It answers the question: *who explains what happened?* Accountability surfaces after action—during audits, reviews, regulatory scrutiny, and failure analysis. It is often formal, sometimes contractual, and rarely comfortable.

Responsibility and accountability are frequently linked, but they are not the same. A party may perform work without being accountable for the final result. Another may be accountable for outcomes they did not directly execute.

The third role is *capability*.

Capability describes who is able to execute. It answers the question: *who can do this well?* Capability is grounded in skill, experience, tooling, and capacity. In technical programs, capability is often the most visible attribute and the most easily mistaken for authority.

Capability is purely descriptive.

The fourth role, and the one most commonly misused, is *authority*.

Authority describes who may decide. It answers the question: *who has the right to choose?* Authority exists before action. It operates under uncertainty. Authority accepts obligation for outcomes that cannot yet be fully known.

The person who performs the work is routinely assumed to have the right to decide it. The person with the greatest capability is assumed to be best positioned to choose. The party who will later be held accountable is assumed to have exercised authority earlier. None of these assumptions are structurally valid.

They persist because they are convenient.

The Fleetwood program made this separation explicit from the outset.

Across AMG, GM platform engineering, suppliers, and later Cadillac engineering, responsibility and capability were distributed broadly. Chassis engineers modeled rigidity. Brake specialists specified systems. Interior craftsmen executed to exacting standards. Platform engineers provided deep structural knowledge. Each carried responsibility. Each possessed capability.

None of them were asked to decide what the vehicle was allowed to become.

Authority over identity, tradeoff priority, and acceptable risk was declared separately. It did not emerge from expertise. It was not inferred from performance. It was not earned through tenure or prior success. Capability informed decisions, but it did not grant jurisdiction.

This distinction mattered.

Cadillac engineering would later stand accountable for platform integrity, safety, and regulatory compliance. AMG would be accountable for the vehicle as a manufactured product. Suppliers would be accountable for delivered components. Accountability was real, explicit, and unavoidable.

Authority, however, was not distributed according to accountability or capability.

It was bounded.

This prevented a common failure mode in complex engineering programs: decisions made by default simply because someone *could* make them. No specialist—regardless of reputation or experience—was allowed to let their domain dominate the system. Expertise constrained choices. It did not select them.

When responsibility, accountability, capability, and authority are blurred, systems accelerate quickly. Work multiplies. Decisions proliferate. Outcomes are produced. Yet no one can later state clearly who had the right to choose any particular path.

The system is active.

It is not governed.

This confusion is especially acute in technical environments. Skilled individuals naturally accumulate influence. High-capability teams are trusted implicitly. Over time, execution becomes indistinguishable from authorization. Decisions are made because they can be made, not because they should be made.

When outcomes are good, this ambiguity is tolerated. When outcomes are bad, it becomes unresolvable.

Postmortems identify responsibility and assign accountability, but authority cannot be located. Everyone participated. No one decided.

This is not a failure of people.

It is a failure of structure.

Clear governance begins with separation.

A system must be able to state—without ambiguity:

- who performs the work
- who answers for outcomes
- who is capable of execution

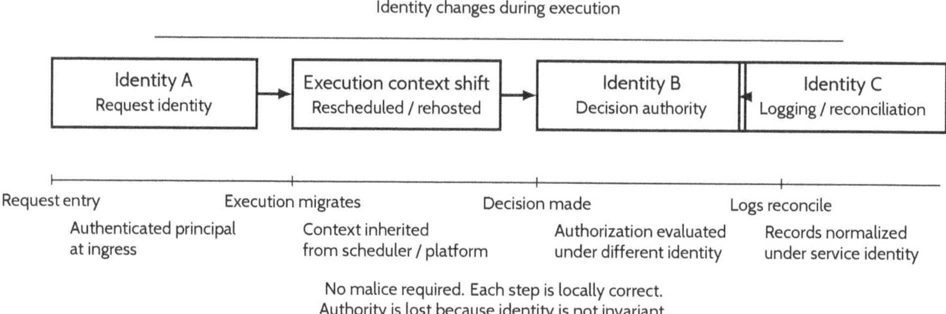

Figure 8.1: Identity drift during execution: a request can enter, execute, decide, and be recorded under different identities. Each component behaves correctly, yet authority dissolves because responsibility is not invariant.

- and who has the right to decide

These roles may coexist within the same individual. They often do. But they must be distinguishable, or authority dissolves into activity.

Until this separation is explicit, any discussion of jurisdiction, legitimacy, or decision rights is premature.

The system will remain busy.

But it will not be governed.

Why Capability Never Grants Jurisdiction

Capability is often mistaken for permission.

In technical systems, the individuals and teams most capable of executing work frequently acquire informal authority simply by virtue of competence. They know the system best. They move fastest. They solve the hardest problems. Over time, their ability to act is quietly conflated with the right to decide.

This conflation is dangerous.

Being able to do a thing does not grant the right to decide that it should be done.

Capability describes what is possible. Jurisdiction defines what is permitted.

These are different categories.

Technical excellence does not imply mandate. Optimization skill does not confer permission. Speed does not justify scope expansion. None of these attributes establish legitimacy over system direction, identity, or risk posture.

Capability is descriptive. Jurisdiction is normative.

This distinction is routinely violated because high capability feels authoritative. Skilled individuals speak with confidence. Their solutions work. Their recommendations are often correct within their domain. Over time, resistance to their decisions feels unnecessary—or even irresponsible.

The system begins to defer.

This is how authority migrates without ever being declared.

The fallacy appears in familiar forms:

- "The smartest engineer decides."
- "The fastest team decides."
- "The best-performing system decides."

Each statement sounds efficient. Each sounds meritocratic. Each is structurally unsound.

Intelligence does not grant jurisdiction. Velocity does not define legitimacy. Performance does not authorize expansion.

A system that allows capability to become authority does not remain governed. It becomes opportunistic.

This failure mode is especially pronounced in software systems.

Senior engineers bypass review because they "know what they're doing." High-velocity teams quietly expand scope because they "can handle it." Performance wins are treated as license—proof that constraints can be relaxed, oversight reduced, and boundaries reinterpreted.

The language is subtle:

- "We'll clean it up later."
- "This won't affect anyone else."
- "It's already working."

Each phrase substitutes execution success for permission.

Over time, governance erodes asymmetrically. The most capable actors gain freedom. Less capable actors remain constrained. Authority becomes unevenly distributed based not on mandate, but on momentum.

This is not meritocracy.

It is drift disguised as efficiency.

In the Fleetwood program, this temptation was deliberately resisted.

AMG possessed extraordinary capability. Its engineers had the skill to chase performance, to select powerplants early, to allow output metrics to dominate integration decisions. They had done so successfully in other contexts.

They were explicitly prevented from doing so here.

Capability informed constraints. It did not determine identity. Specialists were listened to carefully, but they were not granted jurisdiction over tradeoffs outside their domain. Optimization skill did not override declared character. Speed of execution did not justify deviation.

This restraint was structural, not cultural.

It ensured that no part of the system could expand simply because it was strong. Power was required to submit to authority, not the other way around.

This distinction matters because capability is seductive. It feels earned. It feels efficient. It feels safe—especially when outcomes are good.

But systems governed by capability rather than jurisdiction fail in predictable ways. They become brittle. They accumulate hidden risk. They privilege local optimization over global intent. They move quickly until they cannot stop.

Authority exists to prevent this.

Jurisdiction defines who may decide when capability must be constrained. It establishes boundaries that performance cannot cross, regardless of how impressive the results appear.

Without this separation, systems do not remain excellent.

They merely remain busy.

Capability enables action. Authority governs it.

When those roles are confused, the system accelerates toward outcomes it never explicitly chose—and no one is prepared to answer for.

Why Expertise Is Advisory, Not Authoritative

Expertise is indispensable.

Any system that dismisses expertise will fail quickly and loudly. Complex engineering, whether mechanical or software-based, cannot be governed by ignorance. Specialists exist because reality is deep, nonlinear, and unforgiving. Their knowledge constrains fantasy. Their experience reveals failure modes before they are encountered.

But expertise does not decide.

This distinction is essential, and often resisted.

Expertise informs decisions. Expertise constrains options. Expertise explains consequences.

Expertise does not select outcomes.

The confusion arises because expertise speaks with confidence. Specialists understand tradeoffs in ways others do not. They can predict secondary effects, articulate risks precisely, and often identify the technically "best" solution within their domain. Over time, this clarity begins to feel like authority.

It is not.

Experts explain what **will** happen if a choice is made. Authority accepts responsibility for *making* the choice.

These are different obligations.

Expertise is fundamentally descriptive. It describes reality as it is likely to unfold. Authority is normative. It decides which outcomes are acceptable, which risks will be taken, and which tradeoffs will stand. Expertise narrows the decision space; authority chooses within it.

This separation is what allows systems to benefit from deep knowledge without being ruled by it.

When expertise is allowed to become authoritative, systems fail in predictable ways. Decisions begin to optimize for technical elegance rather than intent. Local excellence overwhelms global coherence. Specialists solve problems that matter deeply within their domain while quietly violating constraints elsewhere.

The system becomes impressive and incoherent at the same time.

This is not a failure of experts. It is a failure to constrain them.

In the Fleetwood program, expertise was treated with respect—and limits.

Specialists across AMG, GM platform engineering, suppliers, and later Cadillac engineering informed nearly every constraint. Chassis engineers defined rigidity requirements. Brake experts specified duty cycles that would never be exceeded. Materials specialists constrained weight targets. Interior craftsmen defined tolerances for silence and comfort.

Their expertise shaped the space of permissible solutions.

It did not determine the identity of the vehicle.

Authority over character, tradeoff hierarchy, and acceptable risk was declared elsewhere. Specialists were not asked to decide what mattered most; they were asked to explain what would happen if certain paths were chosen. Their role was to make consequences visible, not to choose which consequences the system would accept.

This distinction preserved coherence.

Experts could argue forcefully without needing to win. Their credibility did not depend on asserting control. Authority could listen without being captured. Disagreement sharpened understanding rather than fragmenting direction.

This balance is rare—and necessary.

In software systems, its absence is common.

Architects become de facto governors. Security experts veto features indefinitely. Performance specialists delay releases without mandate. Domain experts override product intent in the name of correctness. Each believes they are protecting the system.

Often, they are.

But when expertise selects outcomes without declared authority, responsibility becomes ambiguous. Decisions feel technical rather than moral. Tradeoffs are framed as necessities rather than choices. When consequences arrive, no one can say who accepted them.

Authority has been replaced by argument.

This inversion is as dangerous as its opposite. Authority without expertise is negligence. It ignores reality, underestimates risk, and invites failure through ignorance.

But expertise without authority is equally dangerous. It fragments intent, privileges local optimization, and allows systems to drift under the quiet rule of whoever speaks most convincingly.

The engineering truth is simple and non-negotiable: **Expertise without authority is counsel. Authority without expertise is negligence.**

Healthy systems require both—and require them to remain distinct.

Experts must be empowered to explain consequences clearly and without dilution. Authority must be obligated to hear them. But the final decision must rest with those who will answer for the outcome, not with those who merely understand it best.

This is not a hierarchy of intelligence.

It is a division of responsibility.

Only when this division is respected can systems remain both intelligent and governed.

Ownership Does Not Imply Correctness

Ownership is often mistaken for authority.

In long-lived systems, the individuals or teams who created something frequently retain informal control over it long after their original mandate has expired. The logic feels intuitive: *I built it, therefore I understand it; I understand it, therefore I should decide its future.*

This logic is wrong.

Creating a system does not grant permanent authority over it.

Ownership is historical. It describes what happened. It acknowledges contribution. It records effort and intent at a moment in time. Authority, by contrast, is situational. It exists in the present, under current conditions, risks, and obligations.

Past contribution does not grant future jurisdiction.

This distinction is uncomfortable because it challenges a deeply held sense of fairness. Builders feel a legitimate connection to what they create. Their knowledge is real. Their effort is undeniable. Their experience with early tradeoffs often exceeds that of anyone who comes later.

None of this establishes decision rights.

Historical contribution explains *how* a system came to be. It does not determine *what* the system is allowed to become.

The danger arises when ownership hardens into veto power. When original authors assume enduring authority, systems lose the ability to adapt deliberately. Decisions are filtered through precedent rather than obligation. Change is resisted not because it is unsafe, but because it challenges authorship.

The system becomes custodial rather than governed.

Past correctness is often used as justification. *It worked then. We made the right call before. This design has held up.* These statements may all be true. None of them guarantee correctness under new conditions.

Correctness is contextual.

What was optimal at one scale may be dangerous at another. What was safe under earlier assumptions may be fragile under new constraints. Systems evolve. Environments change. Requirements shift. Authority must move accordingly.

When it does not, systems calcify.

This failure mode is especially visible in software.

Original authors retain control over architectures indefinitely. "Legacy owners" block change because they understand the code best. Design decisions made under obsolete constraints are defended as immutable truths. Review becomes ceremonial. Governance defers to history.

The result is predictable.

Innovation slows. Risk accumulates. The system becomes increasingly dependent on the continued presence of its creators. When they leave, authority collapses entirely, because it was never transferred—only deferred.

Ownership replaced governance.

In such systems, correctness becomes temporal rather than structural. Decisions are judged by when they were made, not by whether they are appropriate now. Authority becomes an artifact of seniority rather than responsibility.

This is not respect for experience.

It is abdication.

The Fleetwood program explicitly rejected this pattern.

Historical knowledge mattered deeply. Prior AMG work informed constraints. Cadillac platform history shaped structural decisions. Lessons from earlier programs were surfaced and debated. But no past contribution—by AMG, Cadillac, or any supplier—granted enduring authority over the vehicle's identity or tradeoffs.

Authority was assigned based on current obligation, not historical credit.

This allowed the program to honor experience without being ruled by it. Past correctness informed present constraints, but it did not determine present choices. The vehicle was not required to remain consistent with what had worked before if doing so violated declared intent.

This discipline preserved adaptability.

Authority remained anchored to who would answer for outcomes now—not who had been right before.

The engineering truth is simple:

Ownership is historical. Authority is situational.

Confusing the two is how systems become frozen in time—maintained by those who remember them best, but governed by no one at all.

Only when ownership is acknowledged without being allowed to rule can systems remain both respectful of their past and responsible for their future.

Why Markets and Systems Cannot Decide

When authority becomes uncomfortable, it is often deferred.

Not denied. Not challenged. Simply handed off to something impersonal and ostensibly objective. Two deferrals appear again and again in complex systems:

- "The system will decide."
- "The market will decide."

Both are abdications.

They are appealing because they promise resolution without responsibility. They suggest that truth will eventually surface if enough behavior is observed, enough data is gathered, or enough outcomes are allowed to compete. They imply that decision is something that happens *to* a system, rather than something imposed *upon* it.

This belief is incorrect.

Why Systems Cannot Decide

Systems do not decide. They adapt.

A system optimizes for equilibrium, not intent. It responds to forces placed upon it, reinforcing behaviors that reduce instability and suppressing those that increase it. Feedback loops dampen oscillation. Control mechanisms seek local minima. Over time, the system appears orderly.

Order is not authority.

A system can stabilize around behavior that violates every declared goal. It can become predictable while being wrong. It can function smoothly while accumulating latent failure.

In physical engineering, this is well understood. A chassis will flex until the load paths find a compromise. A drivetrain will absorb torque wherever resistance is lowest. A brake system will fade gradually, compensating just enough to avoid immediate collapse. None of these outcomes are decisions. They are accommodations.

Left ungoverned, the system does not ask what *should* happen. It answers only what *can* happen.

The Fleetwood project explicitly rejected this.

AMG did not allow the vehicle to "settle" into its character through iteration. Rigidity was not discovered after testing; it was declared before construction. Braking margin was not tuned until it felt sufficient; it was specified so it would never need to be questioned. Silence under load was not evaluated as an emergent property; it was enforced as a constraint.

The system was not permitted to decide what was acceptable.

Because if it had been, the result would have been predictable: a vehicle that functioned, passed tests, and violated intent quietly.

This is the fundamental error behind "the system will decide." Systems converge. They do not govern.

Why Markets Cannot Decide

Markets do not decide either. They select.

Markets reward outcomes after the fact. They amplify what survives. They punish what fails. They operate without memory, without intent, and without responsibility. They do not care *why* something works, only that it does.

Selection pressure is not governance.

Markets are extraordinarily effective at producing winners. They are equally effective at externalizing harm. They reward speed over caution, growth over safety, and short-term performance over long-term correctness. Correction arrives only after damage has occurred.

This is why regulated industries do not defer authority to markets.

Aviation does not wait for market failure to discover unsafe aircraft. Medicine does not allow therapeutic harm to "correct itself" through competition. Financial systems do not rely on insolvency as a learning mechanism.

Governance exists precisely because markets arrive too late.

The Fleetwood could never rely on market correction. There was no market. No adoption curve. No iterative feedback loop. There would be no second chance, no volume-based learning, no correction through competition.

If the vehicle failed, it would fail completely.

This forced clarity.

AMG could not justify decisions by precedent. Cadillac could not hide behind platform reuse. Suppliers could not rely on industry norms. There was no diffusion of responsibility through adoption metrics or sales figures.

The only accountability path was human.

This is the point markets obscure: they evaluate survival, not permission. They decide what persists, not what was allowed to exist.

Why Neither Can Accept Responsibility

Authority requires three things:

- Attribution — someone can be named
- Accountability — someone can answer
- Consequence — someone can be sanctioned

Systems and markets provide none of these.

A system cannot testify. A market cannot explain intent. Neither can accept liability. When failure occurs, blame dissolves into process descriptions, usage patterns, or external pressures.

This is why phrases like:

- "The algorithm did it"
- "The system learned this behavior"
- "Users rewarded it" are not explanations. They are evasions.

They describe mechanism, not mandate.

In software, this abdication is routine. Teams defer decisions to autoscaling behavior, algorithmic optimization, or user adoption. Performance regressions are justified by growth. Security gaps are tolerated until exploited. Architectural incoherence is excused because "it works."

The system functions. The market rewards it. Authority disappears.

The Fleetwood program refused this disappearance.

Every major decision had a named owner. Someone accepted the braking margin. Someone accepted the rigidity requirements. Someone accepted the risk of an unresolved engine architecture. Someone would answer if the vehicle failed under sustained Autobahn load.

That accountability was not symbolic. It was structural.

The AMG VIN was not branding. It was liability.

The Engineering Truth

Selection pressure is not governance.

Selection occurs after outcomes. Governance exists before them. Confusing the two is how responsibility is quietly abandoned while systems continue to function just well enough to avoid intervention.

This is not a philosophical distinction. It is a practical one.

When authority is deferred to systems, intent erodes. When authority is deferred to markets, harm is normalized. In both cases, decision is replaced by drift.

The Fleetwood could not drift.

It required authority to be declared in advance, because neither physics nor economics would accept responsibility afterward.

That realization closes the last door of abdication.

If authority cannot emerge from data, consensus, performance, time, systems, or markets—then it must be declared.

And when it cannot be declared, the correct behavior is not improvisation.

It is refusal.

That is where the next bridge begins.

Declared Scope: The Boundary of Legitimate Authority

Authority that is not bounded is indistinguishable from coercion.

Authority that is bounded but not declared is indistinguishable from confusion.

Legitimate authority exists only where scope is explicit.

This is the final distinction required before construction can begin.

Up to this point, the argument has eliminated every false source of authority: data, consensus, performance, time, systems, markets, ownership, capability, and expertise. What remains is authority as a deliberate human act. But declaring authority is not enough. Authority must also declare its limits.

Jurisdiction is not total.

Every legitimate act of authority must answer two questions simultaneously:

- What does this authority cover?

- What does it explicitly not cover?

Without the second answer, the first becomes dangerous.

Why Authority Must Declare Its Boundaries

Authority is power exercised in advance of outcomes. That is what makes it necessary—and what makes it risky.

When authority is asserted without boundary, it expands by default. Decisions spill outward. Responsibility diffuses inward. The authority-holder begins to answer questions they were never meant to answer, while others stop answering questions they are still responsible for.

This is how authority curdles into tyranny—not through malice, but through omission.

Conversely, when boundaries exist but authority is absent, systems fracture. Multiple actors make overlapping decisions. Responsibility becomes ambiguous. Conflicts are resolved through escalation, politics, or inertia rather than intent.

This is chaos—not loud chaos, but bureaucratic chaos: endless meetings, duplicated work, contradictory optimizations.

Legitimate authority lives between these two failures.

It is declared. It is bounded. And it is revocable.

subsectionDeclared Scope in the Fleetwood Program

The Fleetwood project treated scope as a first-order engineering concern.

AMG did not claim authority over everything. It claimed authority over specific domains:

- vehicle identity

- performance envelope

- safety margins

- integration decisions

- final acceptance of risk

Just as importantly, it explicitly did *not* claim authority over:

- Cadillac brand history beyond declared constraints
- supplier internal processes
- GM platform decisions outside the agreed scope
- specialist judgment within their domains of expertise

This separation was not accidental. It was protective.

Cadillac platform engineering retained authority over frame characteristics and historical constraints. Specialists retained authority over explaining consequences within their disciplines. Suppliers retained authority over how they achieved contractual outcomes.

But when tradeoffs crossed domains—when decisions affected identity, safety, or integrated behavior—authority converged. Someone was allowed to decide. Everyone else was allowed to disagree, advise, and warn, but not to veto.

That clarity prevented two failure modes:

- unilateral dominance
- collective paralysis

No one wondered who could decide, and no one wondered who could not.

Boundary as the Opposite of Tyranny

Authority becomes illegitimate when it cannot say "this is not mine to decide."

Unchecked authority absorbs responsibility it cannot competently hold. It silences expertise. It overrides local truth. It mistakes control for correctness.

Bounded authority does the opposite.

It creates space for:

- expertise to operate freely
- responsibility to remain attached
- disagreement to surface safely

It also makes refusal possible.

When scope is declared, it becomes legitimate to say:

- this decision is outside jurisdiction

- this action is prohibited

- this uncertainty cannot be resolved here

Without scope, refusal looks like obstruction. With scope, refusal is compliance.

This distinction is foundational to safety-critical engineering, even when it is not named explicitly.

Software's Failure to Declare Scope

Software systems routinely fail here.

Authority is often implied rather than declared. Teams assume ownership boundaries based on code proximity, velocity, or historical contribution. Decision rights expand quietly as systems grow. No one formally defines:

- who may change behavior under load

- who may trade cost for performance

- who may accept security risk

- who may expand blast radius

As a result:

- service teams make platform decisions

- platform teams impose product behavior

- SRE teams inherit authority without mandate

- architects govern without accountability

When incidents occur, the postmortem reveals the truth: no one knew where authority stopped.

Modern software language gestures at scope — *service ownership*, *blast radius*, *domain boundaries* — but often stops short of authority. Teams know what they operate, but not what they are permitted to decide.

That gap is where failure hides.

Authority Requires Declared Limits

Authority is not legitimate because it exists.

It is legitimate because it is bounded, understood, and accepted by those it governs.

This is why regulators and auditors care less about who made a decision than whether that person was authorized to make it, and whether the scope of that authority was explicit.

Undeclared authority cannot be reviewed. Unlimited authority cannot be trusted. Authority without boundaries cannot be safe.

The Fleetwood project survived complexity not because authority was centralized, but because it was scoped. Everyone knew where decisions ended, where advice began, and where refusal was permitted.

That clarity is what made construction possible.

Without it, the vehicle would have been built under assumption. With it, the vehicle could be built under obligation.

Only once authority is declared *and bounded* does it become legitimate to proceed.

And only then does refusal become meaningful.

That is where the next bridge leads.

Authority Must Be Grantable and Revocable

Authority that cannot be revoked is not authority.

It is capture.

This distinction is essential, because without it, everything argued so far risks being misread as permanence, hierarchy, or dominance. Legitimate authority is none of those things. It is temporary. It is conditional. And it exists only so long as it continues to serve the system it governs.

Authority is assigned, not intrinsic.

No individual, role, or organization *possesses* authority by nature. Authority is granted by a system that recognizes the need for decisions to be made under uncertainty, and it is granted with the explicit understanding that it may later be withdrawn.

This is not a philosophical position. It is a structural necessity.

Why Authority Must Be Grantable

Authority exists to solve a specific problem: the need to decide when multiple plausible paths exist and delay carries cost or risk.

That need is situational.

The moment authority is treated as inherent—attached to a title, an individual, or a past achievement—it stops being a tool and starts being a claim. Decisions become defensive. Challenges are interpreted as threats. The system begins to optimize for preservation of authority rather than correctness.

Grantable authority avoids this failure.

When authority is explicitly granted, everyone involved understands:

- why it exists
- what problem it is meant to solve
- what conditions justify its presence

Authority becomes functional rather than personal.

Why Authority Must Be Revocable

No decision-maker is infallible. No context remains static. No system stays within its original constraints indefinitely.

Authority that cannot be withdrawn persists past its usefulness.

This is how systems drift into failure quietly. Decisions continue to be made under assumptions that no longer hold. Signals that should trigger reevaluation are rationalized away. The authority-holder becomes insulated not by malice, but by momentum.

Revocation is the safety mechanism.

It is how systems correct governance without requiring collapse. It allows authority to be reallocated when conditions change, competence erodes, scope expands, or risk increases beyond the original mandate.

Without revocation, authority becomes self-justifying.

That is capture.

Capture Is the Enemy of Trust

Captured authority is authority that no longer answers to the system.

It cannot be questioned meaningfully. It cannot be constrained. It cannot be removed without disruption.

Captured authority produces predictable pathologies:

- escalation of commitment
- suppression of dissent
- redefinition of success to match outcomes
- reinterpretation of failure as inevitability

Most catastrophic failures do not occur because authority was abused. They occur because authority outlived its legitimacy.

The Fleetwood Discipline

The Fleetwood program treated authority as conditional from the outset.

AMG's authority was real—but not absolute. It existed to define identity, margins, and integration. It did not exist to override physics, dismiss expertise, or suppress constraints surfaced by others.

Just as importantly, AMG's authority could have been withdrawn.

Had the program violated declared boundaries—brand character, safety margins, or integration discipline—its legitimacy would have collapsed. Authority was sustained not by entitlement, but by continued alignment with declared intent.

That possibility mattered.

Because when revocation is possible, authority behaves differently. Decisions become more careful. Listening improves. Refusal becomes legitimate. Overreach becomes visible early.

Authority that knows it can be withdrawn governs differently than authority that assumes permanence.

Software's Failure to Revoke

Software systems routinely fail here.

Authority accretes silently:

- architects become permanent arbiters

- platform teams expand mandate without review

- SRE teams inherit risk without consent

- "temporary" decisions become fixed forever

Because authority is never formally granted, it is never formally revoked.

Instead, it persists until something breaks badly enough to force reorganization. Governance changes arrive as trauma, not adjustment.

This is not agility. It is avoidance.

Revocation is not instability. It is hygiene.

The Engineering Truth

Authority that cannot be revoked cannot be trusted.

Trust requires the possibility of withdrawal. Without it, authority becomes an object to be defended rather than a function to be performed.

Legitimate authority is:

- assigned deliberately

- bounded explicitly

- exercised conditionally

- withdrawn when misaligned

Anything else is not governance.

It is capture disguised as continuity.

With this final constraint in place, authority becomes safe enough to build upon—and fragile enough to be respected.

That is the condition required for the next bridge.

Because when authority can be granted, bounded, and revoked, refusal becomes meaningful.

And only then can construction proceed without pretense.

Closing: Jurisdiction Is Declared, Not Discovered

Authority does not emerge from intelligence.

It does not emerge from performance. It does not emerge from consensus. It does not emerge from longevity.

Each of these can inform authority. None of them can create it.

Throughout this chapter, every familiar substitute for jurisdiction has been stripped away. Intelligence explains possibilities but does not grant permission. Performance validates outcomes but does not retroactively justify decisions. Consensus smooths conflict but does not establish mandate. Longevity creates confidence, not entitlement.

What remains is unavoidable.

Authority arises only when it is explicitly assigned and when the obligation that accompanies it is accepted.

This is why authority feels uncomfortable in engineering cultures that prize optimization, evidence, and emergent behavior. Declaration is not empirical. It is normative. It is an act taken in advance of proof, in full awareness that it may later be judged harshly.

That is precisely why it matters.

A system can operate indefinitely without authority. It can function. It can optimize. It can even appear successful. But without declared jurisdiction, it cannot be governed. When failure occurs—and it always does—responsibility dissolves into mechanism, incentives, or history.

Declared authority prevents that dissolution.

It creates a place where judgment can land. It makes refusal legitimate. It makes construction ethical rather than accidental. And it makes review possible—by peers, by regulators, by history.

This is why jurisdiction must be declared before building begins.

Not because certainty has been achieved. Not because consensus has formed. But because someone has agreed to be accountable for what will follow.

Authority is not something you uncover in a system. It is something you declare—and then agree to be judged by.

Only after that declaration does it become legitimate to proceed.

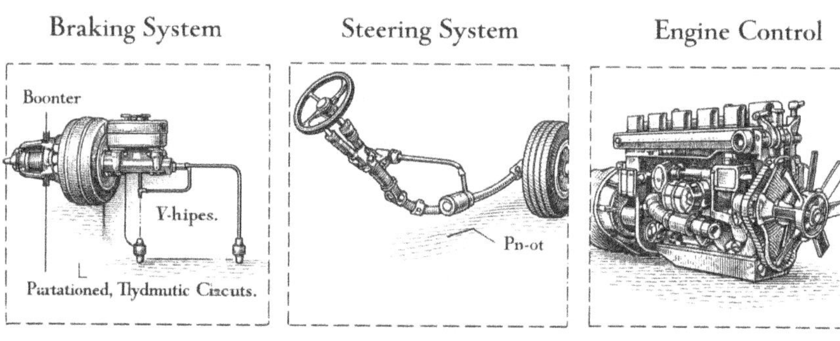

Authority must be declared, scoped, and revocable.

Sanctioned Stillness

When Action Becomes Illegitimate

Action is often treated as neutral.

In many engineering cultures—particularly those shaped by delivery pressure—movement is assumed to be beneficial by default. Progress is measured by activity. Momentum is celebrated. Delay is framed as failure. Stillness is treated as weakness.

This assumption is wrong.

Action without authority is not progress. It is exposure.

When jurisdiction is unresolved, proceeding does not reduce uncertainty—it converts it into ungoverned risk. Decisions made without authority do not remain provisional. They harden. Interfaces form around them. Dependencies accrete. Reversal becomes expensive. What began as motion becomes commitment.

At that point, the system has moved—not forward, but out of control.

Proceeding under unresolved jurisdiction is not a temporary measure. It is a failure state.

This is difficult to accept because action feels responsible. It signals engagement. It reassures observers. It creates the appearance of leadership. But none of those properties confer legitimacy. Without authority, action is indistinguishable from improvisation, regardless of intent.

Improvisation is often mistaken for courage.

The distinction matters.

Courage operates within authority. It accepts consequence on behalf of the system. It acts knowing who will answer if the decision proves wrong. Improvisation, by contrast, replaces authority. It acts *because* authority is absent, not because it is granted. The consequences do not disappear—they are displaced.

Risk does not vanish when authority is missing. It moves downstream. It shifts to users, operators, future teams, or adjacent systems that did not consent to bear it.

Improvisation feels brave because it produces motion under uncertainty. But bravery is not the same as responsibility. Courage carries obligation. Improvisation exports it.

This is why movement without control is not progress.

In governed systems, there are conditions under which action is prohibited—not discouraged, not delayed, but illegitimate. Authority absence is one of those conditions. When no one is permitted to decide, the correct behavior is not to "keep things moving," but to stop.

This is not paralysis. It is containment.

Containment preserves option space. It prevents irreversible commitments from being made under false legitimacy. It protects the system from silently accepting decisions that no one was authorized to make.

The Fleetwood project recognized this explicitly.

When the authority boundary around the engine was unresolved, work did not continue "around" it. No substitute decisions were made to maintain momentum. No placeholders were installed to allow adjacent systems to advance. The absence of authority froze progress by design.

This was not caution. It was discipline.

Allowing construction to proceed under unresolved jurisdiction would have created the illusion of progress while transferring risk to later stages, where reversal would be costly and politically difficult. By refusing to move, the project preserved legitimacy.

This is the inversion most readers must internalize:

Movement is not evidence of responsibility. Stillness is not evidence of failure.

Only action taken under declared authority can be called progress. Everything else is motion without mandate.

Once that distinction is accepted, refusal ceases to feel obstructive. It becomes a required system behavior.

And only systems that can refuse to act deserve to be built upon.

Stillness and Refusal as Load-Bearing Safety

Some systems never stop.

They continue to process requests, accept input, propagate decisions, and advance state regardless of circumstance. This behavior is often praised. Availability is treated as virtue. Responsiveness is treated as reliability. Progress is assumed to be safety.

This assumption is false.

Systems that always proceed are uncontrolled.

A system that cannot refuse to act has no meaningful boundary between permitted behavior and prohibited behavior. Every input becomes actionable. Every condition becomes acceptable. The system may continue to function, but it does so without the ability to distinguish between what should occur and what merely can.

At that point, the system is no longer governed. It is reactive.

This is not resilience. It is capture.

Systems that must act are already captured—by load, by incentives, by automation, or by the absence of authority. They respond because nothing in their design allows them not to. Action becomes compulsory rather than deliberate. Behavior becomes a consequence of pressure rather than intent.

In such systems, refusal requires justification. Action does not.

That inversion is the danger.

Safety does not require that every action be justified. Safety requires that refusal not require justification. A safe system must be allowed to say "no" by default under defined conditions. It must be able to halt execution without apology, escalation, or narrative defense beyond the fact that constraints have been violated.

This is why refusal must be engineered.

A refusal mechanism that depends on human heroics is not safety. A pause that requires political capital is not control. A halt that must be defended rhetorically will eventually be bypassed. When pressure rises, systems revert to their defaults—and if the default is action, action will occur regardless of risk.

Refusal must be structural.

When refusal is optional, systems drift toward continuous operation. When refusal is exceptional, it is delayed until damage is unavoidable. When refusal is stigmatized, it disappears entirely.

A system that cannot refuse to act is not safe to build upon.

This is not a warning. It is a theorem.

Stillness is the operational form of refusal.

Stillness is often mistaken for absence—for lack of leadership, lack of analysis, or lack of progress. In reality, stillness is one of the most demanding behaviors a system can exhibit. It requires restraint under pressure, clarity under uncertainty, and discipline in the face of momentum.

Where action is easy, stillness is expensive.

That cost is precisely why it matters.

Stillness preserves option space.

Every action collapses futures. Each decision commits the system to a narrower set of paths. Interfaces form around early choices. Dependencies emerge. Reversal becomes not merely technically difficult, but socially and politically costly. What begins as a "temporary" step becomes permanent by inertia.

Premature construction hardens mistakes.

This is not a moral claim. It is a mechanical one. Early decisions shape load paths, interfaces, and tolerances. Once concrete is poured, steel is welded, or code is deployed, errors are no longer conceptual—they are structural. Correction requires force. Force introduces risk. Risk demands justification that may never be granted.

Stillness interrupts this collapse.

By refusing to act under unresolved authority, the system keeps multiple futures viable. It prevents placeholder decisions from masquerading as progress. It treats unanswered questions as boundaries rather than inconveniences.

This distinction matters because stillness is not patience.

Patience waits. Discipline refuses.

Stillness, when engineered, is not passive. It is active containment. It absorbs pressure without yielding shape. It prevents authority from being assumed implicitly through momentum.

The Fleetwood project demonstrated this discipline clearly.

When authority over the engine could not be established, progress did not continue "around" the gap. No interim drivetrain was selected to allow other systems to advance. No provisional architecture was installed with the promise of later replacement. The uncertainty was not hidden behind temporary solutions.

It was held. That holding was intentional. Any placeholder would have collapsed the very option space the project was trying to protect. A provisional engine would have dictated weight distribution, cooling requirements, structural reinforcement, and character decisions that could not be undone cleanly. By stopping, the project preserved its ability to decide correctly later.

Stillness carried load.

It supported the integrity of future construction by refusing the convenience of early motion. Downstream systems remained flexible rather than being forced to adapt to decisions made without mandate.

This is the crucial connection between refusal and safety.

A system that can refuse to act can remain still under uncertainty. A system that can remain still can preserve intent. A system that preserves intent can be governed.

In software systems, this discipline is rare.

Systems are designed to be always on, always serving, always deploying. Refusal is treated as outage. Pauses are treated as regressions. Kill switches exist but are rarely used. Circuit breakers are installed but tuned never to trigger. The system proceeds until it cannot—and then fails loudly.

This is not safety. It is deferred failure.

Teams continue building around unresolved questions. Placeholder implementations harden into architecture. Temporary workarounds become permanent interfaces. Uncertainty is masked by abstraction rather than resolved by authority.

The result is predictable. Systems become brittle precisely where they needed to remain flexible. Later corrections require disruptive rewrites, migrations, or compensating complexity. The cost of early motion is paid with interest.

Stillness would have been cheaper.

A controlled system must be able to not act. That ability is not hesitation. It is governance. It is the explicit recognition that some conditions do not merely discourage action—they prohibit it.

Absence of authority is one of those conditions.

When no one is authorized to decide, proceeding is not initiative. It is noncompliance. Action taken under unresolved jurisdiction is unauthorized regardless of care, diligence, or good intent. The system has exited its governance envelope.

Optional pauses can be overridden. Prohibited states cannot.

Only systems that can halt without negotiation can be trusted to resume without having already violated their own limits. Only systems that can carry stillness under uncertainty can carry load under operation.

Without refusal, construction is merely motion—impressive, continuous, and fundamentally unsafe.

Stillness is not the absence of progress. It is the prerequisite for progress that can be defended.

Permission to Stop

Stillness is not a failure mode.

It is the only correct response when authority is absent or contested.

Throughout this chapter, one principle has been established without exception: action taken without authority is not neutral. It is illegitimate. When no one is permitted to decide, the system is not waiting to be courageous—it is obligated to stop.

Authority absence mandates stillness.

This is not caution. It is compliance.

Stillness preserves legitimacy by preventing unauthorized decisions from hardening into structure. It protects intent from being overwritten by convenience. It keeps responsibility attached to those who must eventually accept consequence.

Only governed systems deserve construction.

A system that cannot refuse to act cannot distinguish between what is permitted and what is merely possible. It cannot enforce boundaries. It cannot protect itself from momentum. Such a system may function, but it cannot be trusted.

The Fleetwood project demonstrated this discipline by treating refusal as a prerequisite, not an obstacle. Work did not advance until authority was declared. Silence was held deliberately. Progress was suspended not out of hesitation, but out of respect for legitimacy.

This is the final condition required before building can begin.

Construction is not the act of making something move. It is the act of making something legitimate.

Construction begins only after a system has proven it can refuse to move.

Only then does motion become meaningful.

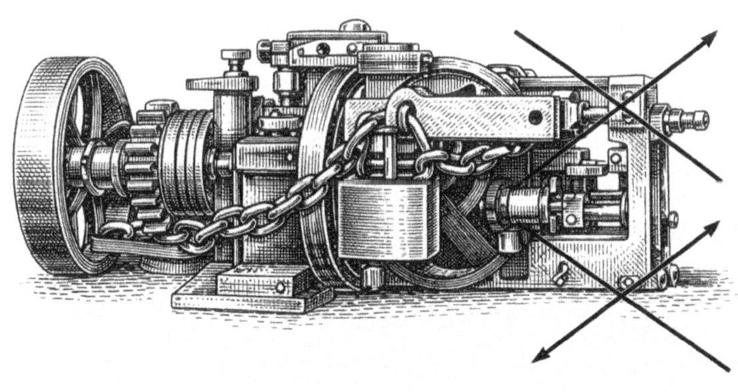

The right to stop is an engineering decision.

10

Jurisdiction Comes Before Capability

Defining the Target Without Designing the Solution

The program did not begin with a design.

It began with a description.

Before there were drawings, before there were models, before anyone asked what engine might fit or how power would be delivered, there was a deliberate effort to define the end state in terms that were immune to negotiation. Not features. Not components. Not aesthetics. Operational truths.

The finished car had to be identifiably Cadillac in character. That meant effortlessness rather than agitation. Torque rather than horsepower. Motion without drama. It had to carry speed the way a large ship carries momentum—quietly, with authority, without constantly reminding its occupants that energy was being expended to keep it moving.

Silence under load was not an aspirational goal. It was an operational requirement.

This was not to be a sports sedan. It was an executive express. That distinction mattered because it constrained every downstream decision. A sports sedan tolerates nervousness. An executive express does not. A sports sedan can trade comfort for immediacy. An executive express must make immediacy disappear.

Speed itself was not controversial. The group agreed early that the car had to be capable of sustained Autobahn velocities—at least 155 miles per hour, or 250 kilometers per hour. Not as a marketing number. Not as a party trick. And not through an artificial governor. The system had to be able to live there.

That requirement immediately reframed the problem. Sustained speed is not about peak output. It is about stability, cooling, braking, rigidity, and refusal. A system that can reach a number briefly but cannot remain composed at that number is not fast—it is fragile.

For AMG, this framing was familiar. Although the company was publicly associated with high-output engines and aggressive tuning, its internal memory ran deeper. AMG was, first, a motorsports company. It remembered machines that delivered power without spectacle—cars that moved decisively rather than theatrically. Someone eventually referenced Briggs Cunningham's Cadillac at Le Mans in the 1950s. Not as a blueprint, and not as nostalgia, but as a reminder: authority does not require nervousness, and performance does not require agitation.

That memory served as a constraint, not an objective.

Just as important as what was defined was what was deliberately deferred.

No engine was selected. No power figures were finalized. No tuning discussions were allowed.

Those conversations were postponed not because they were unimportant, but because they were dangerous. Without invariant success criteria, any discussion of capability collapses into preference. Once preference enters the room, authority leaks out quietly.

The working assumption was simple: if the end state could not be described clearly enough to reject solutions, it was too early to choose one.

This discipline is routine in physical engineering. It is far less common in software.

In software, this discipline is rare.

Very few software projects begin with a fixed, inviolable set of performance demands. Throughput targets are often approximate. Latency expectations are implied rather than stated. Cost envelopes are optimistic.

Failure behavior is discussed abstractly, if at all. Security and accessibility are acknowledged as concerns, but rarely encoded as structural constraints.

Instead of requirements, teams begin with functionality.

As a result, performance work tends not to take the form of validation against known limits. It becomes a benchmarking exercise—an attempt to discover the performance envelope of a system that has already taken shape. Engineers measure what the system can do, rather than confirming that it does what it must.

By the time this discovery begins, dozens—sometimes hundreds—of small decisions have already been made. Architectural choices. Data access patterns. Dependency selections. Concurrency models. Serialization formats. Each decision is locally reasonable. Each decision carries a small performance implication.

Taken together, they form a system whose operational behavior is the sum of assumptions that were never aligned.

Technical debt accumulates quietly in this phase.

Not because engineers are careless, but because they lack a shared definition of what constitutes acceptable performance. Without fixed targets, individual teams optimize for what they can see: convenience, velocity, local correctness. Performance becomes everyone's concern and no one's responsibility.

When conflicts emerge, they are rarely obvious.

One component assumes low latency and frequent calls. Another assumes batching and deferred execution. One team assumes horizontal scalability will absorb inefficiency. Another assumes vertical performance will remain available. Security controls add cost in unexpected places. Accessibility tooling introduces latency where none was budgeted. Each assumption is reasonable in isolation. Together, they collide.

More commonly, performance, security, and accessibility are not actively decided at all.

They are deferred—sometimes explicitly, often implicitly—until they become urgent. At that point, they are no longer design problems. They are rescue operations. Teams add caches. Introduce queues. Layer on controls. Instrument what they can. Each intervention increases complexity. Each increase in complexity narrows the remaining margin.

At that stage, performance is no longer an attribute of the system.

It is a constraint imposed on a system that was never shaped to carry it.

Physical engineering does not afford this luxury. A chassis designed without regard for braking capacity cannot be corrected by larger calipers alone. A frame that flexes under load cannot be fixed by tuning suspension. Operational requirements ignored early do not reappear later as solvable problems; they reappear as limits.

This was the failure mode the Fleetwood program was determined to avoid.

Software systems often begin with functional goals—features, workflows, user stories—while operational requirements are treated as secondary concerns. Performance targets are left elastic. Response times are discussed vaguely. Scalability is assumed to be solvable later. Cost curves are optimistic. Supportability is someone else's problem. Security margins are implied rather than specified. Instrumentation is planned as an add-on.

The result is predictable. Operational characteristics—throughput, latency, cost per transaction, failure modes, attack surfaces—are grafted onto a system that was never shaped to carry them. When performance fails, teams add caches. When security fails, they add controls. When observability is needed, they add tooling. Each addition increases complexity, and each increase in complexity erodes authority.

Physical systems do not survive this pattern.

You cannot bolt braking capacity onto a chassis that flexes. You cannot tune suspension on a frame that lies under load. You cannot add cooling after heat paths are already fixed. Operational requirements are not accessories. They are geometry.

The Fleetwood program treated them as such.

Speed was defined early, not as a number to be reached, but as a condition to be sustained. Silence was defined not as absence of noise, but as absence of disturbance under load. Torque was defined not as output, but as character—how the system moved mass without effort. Cost was not minimized; it was bounded by what would be required to make the system honest. Supportability was implicit: nothing would be hidden that could not later be defended.

Even accessibility—though not named as such—was embedded in the thinking. Instrumentation, diagnostics, and communication pathways were assumed to be necessary, not optional. A system that cannot explain itself under stress is not complete. In physical engineering, telematics, sensors, and feedback loops are designed alongside structure, not appended afterward.

Software too often treats these concerns as afterthoughts.

Operational performance becomes a tuning exercise rather than a design constraint. Security becomes a compliance checklist rather than a margin. Accessibility—understood as the ability to observe, understand, and communicate system state—becomes a tooling decision rather than an architectural one.

The Fleetwood program resisted that drift by fixing the target before allowing solutions to emerge.

By the end of this phase, nothing had been built.

But the boundaries were firm.

The car had to be Cadillac in character. It had to sustain speed without explanation. It had to preserve silence under load. It had to carry margins large enough that braking, cooling, and structure would never become questions. And it had to do all of this without artificial constraints hiding structural weakness.

Only once those truths were fixed did the harder conversations become possible.

One of those conversations stood apart.

The engine.

Everything else in the system could be bounded through inheritance. Braking systems could be oversized using proven components. Suspension behavior could be modeled within known regimes. Electronics, interior systems, and control architectures could be drawn from established authority. Each of those decisions reduced uncertainty.

The engine did the opposite.

It was the heart of the system, and it was the only remaining unknown. Not because engines were mysterious, but because none of the engines available matched the defined character within the defined constraints. Weight distribution, torque delivery, thermal behavior, and brand identity were all in play simultaneously.

This was not a problem that could be solved through documents.

It was not a problem that could be solved through remote collaboration.

When the only unknown is the heart of the system, proximity matters.

And proximity, in this case, was Detroit.

Not because of geography, but because of density. Platform engineers. Drivetrain engineers. Materials specialists. People who knew not just what the drawings said, but why they said it. People who remembered tradeoffs that never made it into documentation. People who could answer questions that were never formally asked.

This recognition did not arrive dramatically. It emerged gradually, as it became clear that resolving the engine question would require more than analysis. It would require trust, shared context, and the kind of informal exchange that only happens when engineers can sit across from each other, draw on napkins, argue over meals, and walk into each other's offices unannounced.

The decision to establish a program office near Detroit was not logistical.

It was architectural.

The system could not move forward until its heart was understood. And understanding would require more than specifications. It would require proximity to truth.

That realization marked the end of definition and the beginning of engagement.

The solutions were still unknown.

But the system was finally constrained enough to ask the right questions.

Rigidity as a Prerequisite for Truth (Distilled)

In high-performance systems, rigidity is often misunderstood.

It is easy to frame stiffness as a performance enhancer—a way to sharpen response or improve precision. In practice, rigidity serves a more fundamental role. It preserves truth.

AMG understood this early. Long before power figures or engine choices entered the discussion, conversations with GM platform engineers returned repeatedly to one issue: torsional rigidity. Not as a tuning parameter, but as a prerequisite. If the structure could not be trusted, nothing attached to it could be trusted either.

This understanding came from experience, not theory. AMG carried institutional memory from the W124 program in the mid-1980s—a vehicle remembered less for headline numbers than for its composure under load. Doors closed cleanly after years of use. Suspension geometry remained coherent. Steering inputs translated consistently. The car did not slowly degrade into vagueness.

That behavior did not come from clever tuning. It came from structure.

The Fleetwood presented a different challenge. Body-on-frame construction offered advantages—ride comfort, isolation, repairability—but it also introduced risk. Unlike a unibody, where loads distribute across a continuous shell, a body-on-frame system creates interfaces. Interfaces flex. And flex absorbs energy.

If the frame deforms before energy reaches the suspension, the suspension lies. If the frame twists under braking, the brakes lie. If torque is absorbed before reaching the axle, the drivetrain lies.

Under those conditions, impressive numbers are still possible. The car may accelerate quickly or corner hard briefly. But the system becomes untrustworthy. Feedback degrades subtly as load increases. Drivers compensate unconsciously—until compensation runs out.

AMG engineers were explicit about this risk. The concern was not that the Fleetwood frame would fail catastrophically, but that it would bend just enough to invalidate everything else. A flexible frame lies quietly. It fails loudly only at the end.

This framing shifted the discussion away from perception and toward measurement. Assumptions were set aside. Engineers compared data rather than reputations. What mattered was not how the structure felt, but how it behaved under torsional load, braking load, and sustained stress.

The goal was not maximal stiffness. It was sufficient stiffness to preserve causality—so energy transferred cleanly through the system rather than dissipating into deformation.

In this sense, rigidity was not about performance. It was about honesty.

When a driver presses the accelerator, response should be proportional. When brakes are applied, deceleration should be predictable. When the suspension carries load, geometry should remain coherent. These are not performance attributes. They are truth conditions.

Software systems face an analogous problem.

Modern architectures are often described as elastic. They scale, adapt, and absorb load. Under normal conditions, this elasticity looks like resilience. Retries mask failures. Buffers delay backpressure. Autoscaling hides inefficiency. Everything appears stable.

But elasticity has a cost.

Just as a flexible frame absorbs mechanical energy, elastic software systems absorb operational truth. Feedback is delayed. Failure signals are softened. Under stress, behavior changes in ways that are difficult to predict and harder to diagnose.

The problem is not failure. It is delayed feedback.

Physical engineers are wary of systems that delay feedback. They prefer structures that reveal limits early, while corrective action is still possible. Rigidity, in this context, is a form of discipline. It keeps cause and effect coupled.

This was AMG's concern. If engine torque twisted the frame, tuning would compensate for structure rather than express capability. If braking loads caused uneven deformation, control systems would chase a moving target. If suspension loads were absorbed inconsistently, handling would become situational rather than repeatable.

These issues do not announce themselves immediately. They emerge over time, under sustained load, often at the margins of performance. By the time they are obvious, options are limited and expensive.

The early AMG–GM dialogue reflected this understanding. Rather than debating solutions, engineers worked to eliminate ambiguity. Measurement replaced intuition. Data became shared language. The objective was not to defend the structure, but to decide whether it could be trusted.

The Fleetwood program treated rigidity as non-negotiable. Before adding power, the structure had to tell the truth. Before tuning behavior, causality had to be preserved. Only then would downstream decisions—brakes, suspension, engine—have meaning.

This was not conservatism. It was discipline.

By insisting on structural honesty early, the program reduced the risk that later performance gains would be illusory. Improvements would reflect real capability, not compensation for hidden deformation.

By the end of this phase, nothing dramatic was visible. But the ground rules were set.

Energy would flow. Deformation would be constrained. Feedback would remain coupled to cause.

Only under those conditions could the system be trusted to reveal its true limits—and only then would it make sense to decide how much power it should carry.

Power Deferred as a Discipline

The question arrived quietly, and once asked, it could not be ignored.

What is a Cadillac engine?

Not historically. Not nostalgically. Not as a displacement figure or a badge. But as an expression of mechanical authority.

A Cadillac engine is smooth. It is balanced. It is not mechanically dominant in the way a hot-rod Lincoln engine once was—urgent, theatrical, demanding attention. A Cadillac engine does not announce itself. It moves mass effortlessly. It produces torque without drama. It sustains motion rather than chasing speed. It is reliable not as a talking point, but as an expectation.

It has mechanical authority.

That authority is not defined by peak horsepower. It is defined by how the engine behaves across time—how it responds under load, how it carries weight, how little it asks of the driver, and how little it demands of the system around it.

Once framed this way, the problem became sharper.

What, among the engines currently produced within the GM portfolio, respected that authority?

The answer, after sustained examination, was uncomfortable.

There were engines that produced impressive numbers. Engines that were efficient, compact, powerful, and modern. Engines that fit platforms cleanly. Engines that performed well in isolation. But none aligned cleanly with the defined character without introducing new uncertainties.

Some were too light in the wrong places, shifting mass forward and altering balance. Some produced power high in the rev range, encouraging behavior inconsistent with effortless motion. Some carried architectural compromises—packaging, vibration, thermal behavior—that would require downstream correction.

Correction was precisely what the program was trying to avoid.

At this stage, torque was expected, but it was not quantified. That restraint was intentional. Quantifying torque too early would have forced premature decisions about structure, cooling, braking, and balance. Without a trusted frame and defined authority boundaries, any number would have been aspirational rather than meaningful.

The risks were well understood.

A powerful engine installed into a structure not prepared to carry it does not merely stress components. It distorts truth. Frame deformation absorbs energy. Suspension geometry compensates silently. Braking behavior becomes inconsistent. The system adapts around weakness instead of expressing capability.

Weight imbalance introduces a different failure mode. Even modest shifts in mass distribution can change how a large vehicle behaves under braking and sustained speed. Corrections cascade. Springs are retuned. Dampers are adjusted. Brakes are oversized further. Each fix addresses a symptom rather than a cause.

Most dangerous of all was the risk of brand character violation.

A Cadillac that feels busy, strained, or eager in the wrong way is not merely flawed—it is wrong. Once that character is lost, it cannot be recovered through tuning. The system may perform well, but it will no longer be authoritative.

These risks led to a deliberate decision: power would be deferred.

Not ignored. Not avoided. Deferred.

This was not indecision. It was discipline.

Rather than selecting an engine and then correcting for its consequences, the program chose to remove as many other unknowns as possible first. Structure, jurisdiction, braking authority, suspension behavior, and collaboration boundaries were fixed before the heart of the system was chosen.

The objective was not to eliminate uncertainty entirely—that would have been unrealistic. The objective was to isolate it.

Only one major unknown would be allowed to remain. Everything else would be constrained.

This approach runs counter to a common pattern in both automotive and software systems.

In software, teams frequently scale before governance is established. Performance tuning begins before ownership is clear. Capacity is added to systems whose authority boundaries are still ambiguous. Power—whether measured in compute, throughput, or concurrency—is applied to unstable architectures in the hope that elasticity will compensate.

Sometimes it does. Temporarily.

More often, it masks structural problems. Latency increases unevenly. Failure modes multiply. Costs climb unpredictably. The system appears strong until it is not.

Physical engineering is less forgiving.

An engine cannot be scaled independently of the structure that carries it. Torque applied to an unprepared frame does not improve performance; it degrades predictability. Power amplifies whatever weaknesses already exist.

The Fleetwood program treated this amplification effect with caution.

By deferring the engine decision, the team preserved optionality. They retained the ability to reject solutions that violated character, balance, or authority. They avoided the sunk-cost trap that often follows early commitment to a powerful component.

This deferral also created space for proper dialogue.

Without a chosen engine, discussions with GM platform engineering remained exploratory rather than defensive. Measurements could be shared without justification. Alternatives could be evaluated without attachment. When the conversation eventually turned toward new block possibilities and historical analogs, it did so from a position of necessity rather than ambition.

Importantly, this deferral did not slow the program.

Work continued on everything that could be known. Braking systems were sized conservatively. Structural reinforcements were evaluated. Suspension architectures were modeled. Jurisdiction was declared. Collaboration matured.

Momentum came not from horsepower, but from clarity.

In hindsight, it can be tempting to frame this restraint as foresight. It was not. It was risk management. The team understood that once power entered the system, it would dominate subsequent decisions. Until the system was ready to tell the truth under load, power would only distort feedback.

Deferring capability is sometimes the most responsible design decision.

It signals confidence rather than hesitation. It reflects an understanding that authority is not established by strength alone, but by the conditions under which strength can be expressed honestly.

By the end of this phase, the engine remained unresolved.

That uncertainty was not a failure of progress. It was evidence of discipline.

The system had been shaped to receive power—but only when power could be trusted to reveal, rather than conceal, the truth.

Only then would the question of what a Cadillac engine should be become answerable.

Jurisdiction Is Declared, Not Discovered

By the time questions of authorship arose, the technical relationship between AMG and GM Platform Engineering was already strong.

That strength did not come from alignment meetings or formal governance. It came from repeated, unglamorous work: engineers comparing measurements, interrogating assumptions, and discovering—sometimes uncomfortably—where their mental models diverged. Over time, a shared language developed. Not consensus, but trust in method. Disagreement became productive rather than political.

Cadillac engineering was eager to re-engage. Cadillac had already had a series of discussions with AMG while the new Seville STS was in development about whether it would make sense to partner with AMG on the car. Complications involving GM's financial leadership and confusion over the final product's brand identity eventually scuttled the relationship.[1] The Cadillac and AMG engineers realigned quickly, not because their brands were similar, but because their instincts were. Both cared about how systems behaved under load. Both preferred uncomfortable truths to comforting abstractions.

That relationship mattered. But it was not enough.

Trust enables collaboration. Jurisdiction enables execution.

This distinction is easy to miss in environments where cooperation is prized. Strong relationships can mask structural ambiguity for a long time. Work progresses. Problems get solved. Decisions are made. But without declared authorship, every decision remains provisional. Authority leaks sideways. Accountability diffuses quietly.

At a certain point in the program, that risk became explicit.

AMG could not proceed as a modifier, advisor, or co-designer. The scope of the work—and the uncertainty still contained within it—required singular authorship. Not because AMG wanted control, but because the system demanded it. Once execution began, disagreements could not be negotiated mid-flight. They had to be resolved before motion.

---

[1]https://www.hagerty.com/media/automotive-history/when-amg-almost-went-american/

That moment marked a transition.

AMG explicitly accepted authorship of the finished vehicle. GM, in turn, released the platform early and stepped back from downstream decisions. This was not abdication. It was clarity. GM retained responsibility for the platform as delivered. AMG assumed responsibility for everything that followed.

That boundary mattered because it could not move.

Jurisdiction, once declared, had to remain fixed. If authority were allowed to drift during execution, every decision would become a negotiation. Velocity would slow. Risk would spread. Responsibility would blur.

The clearest expression of this transfer was not contractual. It was physical.

The bodyshell pull.

Rather than starting with a completed vehicle and undoing it, the car would be removed from the production line at a precise stage—before drivetrain installation, before interior assembly, before decisions that would later need to be reversed. The shell would leave the line intentionally incomplete.

RUF has this relationship with Porsche, Alpina with BMW. This would be a first for AMG. With Mercedes, AMG had already established the pattern of modifying completed cars. Now it would have full responsibility, fixed boundaries, no shared ownership once execution began. This was not experimental governance. German KBA Manufacturer status would be secured in advance of construction.

This was not a convenience. It was a jurisdictional act.

Once the bodyshell left the line, ambiguity left with it. From that point forward, AMG owned the system as a whole. There would be no shared blame if something failed. No quiet escalation back to the platform owner. The act of removal was an acceptance of consequence.

That acceptance was formalized through issuance of a manufacturer VIN.

A VIN is not branding. It is authorship in legal, regulatory, and reputational terms. It says, plainly: this entity stands behind the system. It cannot defer responsibility. It cannot redirect scrutiny. The vehicle is what it is because this organization made it so.

That precedent mattered. GM was not being asked to invent a new model—only to recognize one that already worked. Cadillac engineering, observing the transfer, could see the logic. Jurisdiction was not being seized. It was being declared.

This distinction is subtle but critical.

Jurisdiction is not an organizational artifact. It does not live in reporting structures, steering committees, or escalation paths. It lives at the boundary where decisions stop. It defines where authority ends and where consequence begins.

Systems without clear jurisdiction rarely fail immediately. They fail ambiguously. When something goes wrong, activity increases. Meetings proliferate. Postmortems are written. But responsibility diffuses instead of concentrating. The system learns slowly, if at all.

Modern software provides a familiar parallel.

Many SaaS platforms guarantee uptime—an infrastructure-scoped, binary attribute—while explicitly excluding guarantees around response time, throughput, tail latency, or load isolation—business-scoped attributes that are expensive to own. Providers retain the right to govern load while limiting responsibility for user experience.

The system may be "up." The business may still be down.

In these arrangements, customers bear consequence without jurisdiction. Providers control behavior; customers absorb impact. This is not a technical failure. It is a structural choice—and an unstable one.

A system where one party governs behavior and another absorbs consequence is not engineered. It is improvised.

The Fleetwood program rejected this model explicitly.

By declaring jurisdiction early, AMG made it possible to move quickly later. Risk could be isolated rather than distributed. Uncertainty could be contained rather than spread. Relationships remained strong precisely because boundaries were clear.

The system was not safer because everyone agreed.

It was safer because everyone knew where agreement ended.

That clarity marked the end of definition—and the beginning of execution.

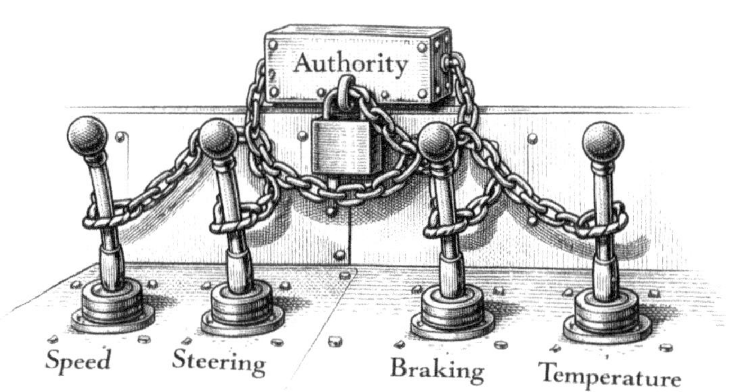

Power without permission is risk.

11

Safety Margins Are Not Waste

If a system requires perfect behavior to survive, it is already unsafe.

Why Brakes Come First

In any system that moves, stopping is not a secondary concern.

It is tempting to treat braking as a response mechanism—a safety system that exists to correct mistakes made elsewhere. In practice, braking serves a more fundamental role. It is the authority that governs motion itself. It determines when movement is allowed to continue and when it must end.

Without that authority, motion is not power. It is risk.

This is why, in the Fleetwood program, braking decisions preceded engine decisions. Not as a hedge, and not as conservatism, but as discipline. The engine remained unresolved by design. Braking behavior could not.

Unknown power was acceptable. Unknown stopping authority was not.

This sequencing mattered because it enforced a hierarchy. Propulsion creates possibility. Braking establishes permission. One without the other produces a system that can accelerate into conditions it cannot safely exit.

Brakes occupy a unique position in the system. They are the last line of defense when everything else fails, but they are also the first system that must never lie. Engines can misbehave and still be managed. Suspensions can flex and recover. Structures can deform within tolerance. Brakes cannot equivocate. If braking authority is compromised — by fade, by thermal overload, by inconsistent response — then every downstream signal becomes suspect.

A system that lies about its ability to stop cannot be trusted to move.

For this reason, the Fleetwood program treated braking not as a performance component but as a truth-preserving one. The goal was not to achieve impressive stopping distances under ideal conditions. The goal was to eliminate the question of stopping altogether. The system must never ask, *"Can it stop?"*

That question had to be answered once, architecturally, and then removed from consideration.

Once removed, something important happens. Cognitive load drops. Engineers stop negotiating with physics. Downstream systems stabilize because they are no longer compensating for uncertainty at the boundary. Authority is preserved precisely because it does not need to be exercised repeatedly.

This principle has a direct analogue in software systems, though it is rarely framed the same way.

Every software system has its own version of brakes. Admission control. Rate limits. Backpressure. Circuit breakers. Kill switches. These mechanisms decide when work is allowed to proceed and when it must be refused. They are not reliability features, and they are not operational tooling. They are authority mechanisms.

Yet in many software architectures, these mechanisms are treated as optional enhancements—something to be added after throughput is optimized, after concurrency is increased, after scale is demonstrated. Capability is applied first. Authority is grafted on later.

The result is familiar.

Systems scale until they cannot. Load increases until queues grow unbounded. Retries amplify contention. Autoscaling hides inefficiency. Monitoring dashboards remain green while users experience delay. Under nominal conditions, everything appears stable. Under stress, behavior changes abruptly.

Just as a flexible frame absorbs energy and delays feedback, elastic software architectures absorb operational truth. They defer failure instead of preventing it. When limits are finally encountered, they arrive without warning and without clarity.

This is the software equivalent of asking, at runtime, *"Can it stop?"*

At that point, authority has already failed.

The Fleetwood program refused this pattern. By selecting braking authority before power was known, the team removed an entire class of failure from the system. Fade would not enter the discussion. Thermal panic would not dictate behavior. Driver hesitation under repeated load would not be a factor. The brakes were not sized to impress; they were sized to be forgotten.

In software terms, this is the difference between systems that rely on perfect behavior and systems that are designed with margin. Perfect behavior assumes ideal inputs, cooperative users, and predictable demand. Margin assumes the opposite. It assumes spikes, contention, misuse, and error—and it plans accordingly.

Systems that rely on perfect behavior instead of margin are already broken.

They may function in demonstrations. They may pass benchmarks. They may meet uptime SLAs. But they do so by negotiating with their limits rather than enforcing them. When those negotiations fail, the failure is rarely contained.

Overspec as Failure-Class Elimination

In many engineering cultures, overspecification is treated as waste.

Excess capacity is framed as inefficiency. Margin is seen as cost rather than protection. Components larger than strictly necessary are assumed to reflect uncertainty or lack of optimization. This perspective is common in organizations that optimize against models rather than behavior.

AMG did not share that view.

Within AMG, overspecification was not additive engineering. It was subtractive. The objective was not to increase performance, but to remove entire categories of failure from the system before they could manifest.

This philosophy was not theoretical. It was practiced.

AMG routinely fitted S-Class braking systems to smaller sedans—not because the vehicles required that level of stopping power under ideal conditions, but because the duty cycles they would experience in reality were neither ideal nor predictable. Sustained high speed, repeated deceleration, heat soak, uneven surfaces, and real drivers did not resemble laboratory assumptions. Larger braking systems ensured that thermal limits never became operational variables.

In more extreme cases, AMG sourced braking components designed for armored vehicles and applied them to standard platforms. The reasoning was consistent. Systems designed to manage mass far beyond what they would ever encounter in service behave calmly when that mass never arrives. Heat remains controlled. Fade never enters the discussion. Pedal feel remains invariant not because it is delicately tuned, but because it is never stressed.

The result is not spectacle. It is consistency.

Braking authority that does not change under load preserves truth. When response remains constant, engineers stop compensating. Suspension tuning stabilizes. Structural behavior becomes easier to measure. Driver confidence becomes implicit rather than managed.

Most importantly, entire classes of questions disappear.

Overspecification eliminates failure modes instead of negotiating with them repeatedly.

This is why larger-vehicle duty cycles were deliberately applied to the Fleetwood's braking system. The goal was not to stop the car once, or even to stop it well. The goal was to ensure that stopping would never require reconsideration—regardless of speed, repetition, or load.

Margin, in this context, is not excess.

Margin is the removal of an entire category of failure.

This distinction becomes clearer when contrasted with how software systems often approach capacity.

In software, "mass" is rarely treated as a first-class concept. Unlike physical systems—where weight, load paths, and thermal behavior are measurable and routinely modeled—software mass is often implicit. CPU utilization, memory pressure, I/O contention, lock behavior, cache churn, and garbage collection are real, quantifiable phenomena, yet they are rarely treated as design constraints.

Not because they cannot be known.

But because curiosity about them is not rewarded, and the skills required to ask the right questions are not consistently taught.

Software teams are trained to produce features, not to understand how those features consume resources under load. The fastest implementation that satisfies functional requirements is rarely the most mass-efficient option. Poor resource usage accumulates quietly, hidden behind abstractions and frameworks that make inefficiency easy and visibility rare.

These inefficiencies become unknown-unknowns.

Not unknown because they are mysterious, but unknown because no one has been incentivized to measure them.

The consequence is familiar. Instead of overspecifying components against known mass, software systems overspecify infrastructure against unknown behavior. Larger instances are provisioned. Autoscaling thresholds are widened. Kubernetes clusters expand horizontally. Safety margins are purchased rather than engineered.

This is compensatory overspecification.

It does not remove failure classes. It masks them.

Higher hosting and operational costs become the price of uncertainty. Capacity is expanded not because demand requires it, but because behavior is poorly understood. Tail latency emerges because resource contention was never modeled. Systems collapse under peak load not because capacity was insufficient, but because the true cost of each request was never known.

When failures occur, the explanation is often the same: we've never seen that before.

That statement is not evidence of rarity. It is evidence of blindness.

AMG's overspec philosophy rejects this posture. It begins with known mass. Known duty cycles. Known worst-case behavior. Components are then selected to exceed those realities by design, not by budget. Once selected, entire classes of failure cease to exist.

The Fleetwood program applied this discipline deliberately. Braking authority was oversized not because the car was expected to be abused, but because the system was not allowed to be surprised.

Overspecification, done with understanding, simplifies systems.

It narrows the space of possible failure. It stabilizes feedback. It preserves authority under stress.

In both physical and software systems, margin is often criticized for what it costs. It is rarely appreciated for what it removes.

AMG understood that the most expensive failures are not the ones that occur—but the ones that require permanent compensation because they were never designed out.

That is not waste.

That is engineering with memory.

Brakes That Can Hold the Engine

By this point in the program, one assumption was no longer debated.

Torque would define the character of the car.

Not peak horsepower. Not redline theatrics. Not transient aggression. The defining trait would be how the system moved mass from rest, how it sustained motion without effort, and how little strain it communicated while doing so. Torque was not a performance metric; it was an expression of authority.

What remained unknown was its magnitude.

That uncertainty was intentional. The exact torque figure was deferred along with the engine itself. But the dominance of torque as a governing force was already accepted. Any future engine solution would be judged primarily by how it delivered torque across the operating range, not by how it performed at the margins.

This framing introduced a non-negotiable requirement for braking.

The brakes could not merely decelerate the vehicle. They had to be capable of holding against the engine's output.

This distinction matters.

Deceleration is a dynamic condition. It assumes motion and removes energy over time. Holding against torque is a static assertion of authority. It requires the braking system to exceed the engine's ability to apply force, regardless of whether the car is already moving. It is the difference between slowing something down and refusing to let it move at all.

That refusal had to be absolute.

This requirement existed before power was known. It existed before displacement was finalized. It existed before cam profiles, compression ratios, or fueling strategies were discussed. The braking system had to be capable of asserting control over whatever power ultimately entered the system.

Control had to exceed capability.

This was not about worst-case scenarios or emergency behavior. It was about hierarchy. In a well-structured system, propulsion is subordinate to control. Power is permitted only insofar as it can be governed. Once that relationship is inverted—once capability exceeds control—the system becomes unpredictable, regardless of how refined its components may be.

Physical engineering enforces this truth ruthlessly.

An engine that can overpower its braking system does not merely threaten safety. It distorts feedback. Drivers hesitate. Engineers compensate. Systems adapt around weakness rather than expressing intent. The vehicle may still function, but it no longer behaves honestly.

This is why the braking system was required to hold the engine, not merely slow the car.

The same hierarchy applies in software systems, though it is often violated quietly.

Many software architectures include rate limits that cannot constrain their producers. Limits exist, but they are advisory. Producers generate work faster than consumers can process it, and the excess is absorbed by queues. Those queues grow. Latency increases. Consumers adapt by retrying, batching, or timing out. The system remains "up," but authority has shifted.

In these systems, control does not exceed capability.

Queues absorb load but never push back. They act as shock absorbers rather than governors. Instead of refusing work when limits are reached, the system accepts everything and defers the cost. Eventually, the cost arrives all at once—through timeouts, cascading retries, or resource exhaustion.

A slowed system that remains responsive preserves authority; a saturated system that stops responding has already lost it.

This is the software equivalent of brakes that can slow a car but cannot hold it on an incline.

The system appears stable under normal conditions. Under sustained pressure, it drifts. Engineers tune consumers. Increase resources. Add buffers. None of these address the root issue: sources are never required to slow down.

Authority has been ceded.

In a properly governed system, producers are constrained by the system's ability to absorb work. Rate limits are enforced, not suggested. Backpressure propagates upstream. When limits are reached, work is refused rather than deferred indefinitely. The system remains responsive precisely because it is allowed to slow.

Control exceeds capability.

The Fleetwood program applied this same principle physically. Braking authority was required to dominate propulsion authority, regardless of how that propulsion was eventually realized. This ensured that when power entered the system, it would be subordinate rather than sovereign.

That subordination mattered.

It meant that future engine discussions could proceed without fear that braking would need to be revisited. It meant that torque could be explored without destabilizing the rest of the system. It meant that authority boundaries would remain intact even as capability increased.

By enforcing this relationship early, the program avoided a common failure mode: systems that grow powerful faster than they grow governable.

The engine would come later.

When it did, it would enter a system already prepared to control it.

That preparation was not dramatic. It was quiet. But it ensured that when capability finally arrived, it would not overwhelm authority.

And that distinction would matter far more than any number printed on a specification sheet.

Consistency Over Peak Performance

Peak performance is easy to demonstrate.

It can be measured in isolation, achieved under controlled conditions, and repeated just long enough to satisfy a test plan. In physical systems, peak numbers are often impressive. In braking systems, they are also largely irrelevant.

Anyone can stop once.

Authority is revealed only through repetition.

For the Fleetwood program, the meaningful test was not how quickly the car could be brought to rest from speed under ideal conditions. It was whether the tenth stop behaved like the first. Whether heat cycles altered response. Whether pedal feel drifted. Whether braking authority degraded quietly as stress accumulated.

If behavior changed, authority was already compromised.

Consistency was therefore not treated as a secondary characteristic. It was the primary requirement. Braking systems were evaluated not for their maximum output, but for their refusal to change under sustained load. Thermal capacity was not a buffer to be consumed; it was margin intended never to be noticed.

This focus on repeatability had a deliberate effect.

When braking response remains invariant, interpretation disappears. Drivers do not adjust behavior. Engineers do not compensate through tuning. Downstream systems—suspension, structure, control logic—are allowed to operate against stable assumptions.

Consistency produces calm.

Calm is not the absence of power. It is the absence of doubt.

The importance of this distinction becomes clearer when viewed through the lens of constrained recovery.

A braking system that stops decisively once but requires extended cooling before it can be trusted again is not authoritative. It forces operators to change behavior after stress. It introduces hesitation. It replaces confidence with calculation.

The same pattern appears in software systems, particularly those that are internally facing and mission-based.

These systems are not judged by engagement or conversion. They govern the ability to work at all. Shipping systems. Inventory lookup. Customer support platforms. Accounting, payroll, compliance, timekeeping. When these systems slow, value creation slows with them. When they stop, it stops entirely.

This is a fundamentally different operating environment from consumer-facing internet services. A consumer application may tolerate degraded experience while remaining nominally "up." An internal system cannot.

When a warehouse cannot ship, when a call center cannot access records, when finance cannot post entries at close, the organization does not degrade gracefully. It stalls.

In these environments, repeatability under constraint matters more than peak capability.

A slowed but responsive system may frustrate users, but work continues. A saturated system that becomes unresponsive removes the ability to generate value altogether. The distinction is not subtle. It is binary.

Studies have suggested that significant portions of the workday can be lost to software friction—systems that are not down, but slow[1]. That time is not idle. It is fragmented. It interrupts flow. It compounds across teams. Over weeks and months, it becomes a structural tax on the organization.

Outages are more visible, but slowdowns are often more expensive.

They erode trust gradually. Users adapt defensively. Workarounds emerge. Manual processes reappear. The system remains present, but its authority weakens.

This is the software equivalent of a braking system that behaves differently after stress. Operators begin to anticipate failure. They leave more distance. They brake earlier. They drive differently—not because the system has failed, but because it can no longer be trusted to behave consistently.

In both domains, the second failure is always worse than the first.

The first failure surprises. It may even be forgiven. The second confirms suspicion. Once trust is lost, operators no longer interact with the system as designed. They adapt around it. At that point, authority has already been forfeited.

This is why validation is not enough.

Software systems frequently pass load tests once. Benchmarks are run under idealized conditions. Performance is demonstrated, recorded, and archived. That success is mistaken for a guarantee.

But validation is a moment in time.

Enforcement is structural.

A system that is validated but not enforced will drift under real load. Resource contention will emerge. Tail latency will grow. Recovery times will lengthen.

---

[1] https://www.prnewswire.co.uk/news-releases/technology-issues-identified-as-a-major-contributor-to-the-uks-productivity-gap-could-cost-uk-plc-35-631234113.html

Operators will compensate. The system may still function, but it will no longer behave predictably.

Predictability is what authority looks like in operation.

In the Fleetwood program, consistency was engineered through margin. Thermal capacity was sufficient not merely to survive repeated stops, but to make repeated stops unremarkable. Recovery time was short because recovery was never required. The system never crossed thresholds that demanded reinterpretation.

In software systems, consistency emerges when resource usage is understood, constrained, and recovered predictably. Not when capacity is simply expanded. Adding resources may improve peak throughput, but it does not guarantee repeatable behavior. In many cases, it extends recovery time and increases coordination overhead.

Just as larger brakes are not installed to stop harder, but to stop the same way every time, well-governed software systems are not designed to be fast once. They are designed to behave consistently under sustained, imperfect conditions.

Consistency removes interpretation.

When behavior does not drift, users do not adapt defensively. When recovery is predictable, systems remain usable under stress. When responsiveness is preserved—even at reduced speed—authority remains intact.

Peak performance impresses.

Repeatable performance earns trust.

And trust, once lost, is far more expensive to recover than any margin ever was.

Safety Margins as Cultural Memory

Safety margins rarely begin as policy.

They are not typically invented in conference rooms or justified in isolation. In engineering organizations that have endured sustained stress, margins emerge differently. They accumulate. They persist. They survive long after the incidents that produced them fade from explicit memory.

Within AMG, margin was not a theoretical construct. It was cultural memory.

AMG's early identity was shaped in environments where systems were subjected to repeated, prolonged stress. Motorsports does not reward peak capability achieved briefly. It punishes degradation mercilessly. Components that perform brilliantly for a few laps but change behavior as heat accumulates are not merely inconvenient—they are untrustworthy.

In that context, failure is not abstract. It is visible, audible, and immediate. Parts fail late rather than early. Small miscalculations surface only after repetition. Fatigue reveals itself not as warning, but as breakage.

Margins grew out of that reality.

They were not added to be cautious. They were added because engineers had already seen what happened when capacity was consumed completely. A brake that faded late in a race. A component that behaved differently after hours under load. A system that passed inspection but failed under repetition.

Each margin represented a lesson that did not need to be relearned.

Over time, those lessons became embedded. Brake sizing reflected not theoretical requirements, but remembered failures. Cooling capacity reflected not average conditions, but the worst sustained loads engineers had already witnessed. Structural reinforcements appeared where fatigue had once emerged unexpectedly.

This knowledge did not always survive as stories.

It survived as structure.

Margins persisted even as teams changed, projects ended, and individuals moved on. New engineers inherited decisions whose origins were not always explained, but whose consequences were understood implicitly. Certain failures simply did not happen anymore.

This is how institutional memory becomes durable.

It does not depend on documentation alone. It is carried forward in default choices, design instincts, and refusal to revisit settled questions. The metal remembers even when people forget.

The Fleetwood program inherited this posture.

Overspecification was not justified purely through analysis. It was justified through lineage. Components were chosen because they had already endured conditions beyond what the program would impose. Larger duty cycles were selected not out of fear, but because smaller ones had failed elsewhere, in other contexts engineers still remembered.

This is not conservatism.

Conservatism narrows ambition to avoid risk. Memory narrows uncertainty to enable ambition. AMG's margins existed so power could be applied later without surprise, not so it could be avoided altogether.

That distinction matters as the system evolves.

Margins that emerge from experience behave differently than margins imposed from policy. They are not optimized away easily. They are defended instinctively. They persist even when pressures mount to reduce cost, simplify designs, or reclaim capacity.

But memory is not permanent.

When the context that produced margin fades, margins themselves can begin to look arbitrary. Oversized components are questioned. Capacity is reclassified as inefficiency. Safety mechanisms are reframed as legacy artifacts rather than earned protections.

At that point, margin does not disappear dramatically.

It erodes quietly.

The next sections will examine what happens when systems—particularly software systems—lose this form of memory, and how safety, once separated from authority, becomes something negotiated rather than enforced.

Here, it is enough to observe this:

Margins exist where engineers have already been hurt.

They do not announce themselves as wisdom. They simply prevent certain failures from ever reappearing.

What happens when that memory is no longer carried forward is a different question—and one the chapter is not yet finished asking.

Software's Quiet Rejection of Margin

In software systems, safety margins are rarely rejected explicitly.

There is no meeting where a team decides that margin is unnecessary, or that limits will be ignored. Instead, margin disappears gradually, displaced by assumptions that feel reasonable in isolation and invisible in aggregate.

Retries will save us. Autoscaling will absorb errors. High availability will cover experience.

Each of these assumptions is defensible on its own. Together, they form a system that functions only when behavior is perfect.

Retries are introduced to handle transient failure. In practice, they often amplify it. A component that is slow or overloaded receives more work precisely when it is least able to handle it. Latency increases. Contention grows. Failure spreads outward, not because retries are flawed, but because they replace refusal with persistence.

Autoscaling is introduced to manage variable demand. It does increase capacity. It also delays feedback. Systems appear stable longer than they should. Inefficiencies remain hidden behind expanding resource pools. When limits are finally reached, they are reached everywhere at once.

Availability targets—"five nines," "four nines," contractual uptime—are treated as proxies for safety. But availability measures presence, not behavior. A system can be available and unusable. It can respond and still fail to serve. Experience degrades long before uptime is threatened.

In these environments, margin is replaced by hope.

Hope that failures will be rare. Hope that demand will resemble forecasts. Hope that elasticity will compensate for inefficiency.

When systems behave well, these hopes go unchallenged. When they behave poorly, the response is rarely to restore margin. Instead, more layers are added. More retries. More capacity. More contracts. Each layer postpones consequence while increasing complexity.

This is not negligence.

It is incentive-aligned behavior.

Software teams are rewarded for shipping features, reducing visible cost, and meeting contractual targets. Safety margins are invisible when they work and blamed when they appear expensive. Over time, mechanisms that exist solely to prevent rare but severe failures are reclassified as overhead.

Margin erodes quietly.

The result is a class of systems that are fragile in a specific way. They perform acceptably as long as inputs are well-behaved. They degrade unpredictably when those assumptions are violated. Operators adapt defensively.

Users change behavior. Workarounds proliferate.

The system remains "up," but its authority weakens.

Unlike physical systems, software can hide this loss for a long time. Logs rotate. Incidents are renamed. Infrastructure grows. The cost of ignorance is paid monthly rather than immediately. Failure does not arrive as a single event. It accumulates as latency, friction, and distrust.

Eventually, the system encounters a condition it cannot absorb.

At that point, the absence of margin becomes visible—not as a single fault, but as cascading loss of control. Retries overwhelm dependencies. Queues saturate. Autoscaling reacts too slowly or too broadly. Contracts define responsibility, but do not restore function.

This outcome is often described as unexpected.

It is not.

Systems that rely on perfect behavior instead of margin are already broken.

They are broken not because they fail often, but because they have no mechanism to enforce limits when behavior deviates from expectation. They negotiate with load rather than governing it. They adapt around stress instead of refusing it.

This rejection of margin is rarely malicious. It is quiet, incremental, and normalized. Each decision appears reasonable. The system as a whole becomes brittle without anyone declaring it so.

The next section will address what happens when safety is treated not as protection from failure, but as the means by which authority is preserved when failure inevitably arrives.

Here, it is enough to name the pattern.

Margin was not removed.

It was never defended.

Safety as Authority Preserved

By the time the braking system was finalized, something important had already happened.

Questions had been removed.

This was the real function of overspecification. The brakes were not oversized to improve a number, impress a driver, or future-proof an unknown engine. They were oversized to eliminate an entire category of uncertainty from the system.

Once that uncertainty was gone, everything else became simpler.

Engineers no longer needed to ask whether repeated high-speed deceleration would alter behavior. They did not need to model fade scenarios or debate cooling margins. They did not need to revisit brake selection every time power was discussed. The braking system had exited the design conversation entirely.

That absence mattered.

When questions disappear, cognitive load drops. Engineers stop carrying conditional logic in their heads. Decisions downstream stabilize because they are no longer compensating for potential weakness at the boundary. Suspension tuning can assume invariant braking behavior. Structural analysis can proceed without hedging. Control systems can rely on consistent feedback.

Authority, in this sense, is not asserted. It is preserved.

The system never negotiates with physics. It does not ask whether it can stop, or under what circumstances stopping might change. It does not require vigilance, interpretation, or restraint from the operator. Its limits are enforced structurally, not behaviorally.

This preservation of authority is what makes power possible later.

When safety margins are real—when they are embodied in the system rather than assumed—capability can be added without fear. Torque can increase without destabilizing control. Speed can be sustained without introducing doubt. Stress does not force reinterpretation.

Trust emerges not because the system is gentle, but because it is honest.

The same pattern applies beyond braking.

In any engineered system, authority survives only where limits are enforced before they are tested. When safety is treated as an operational concern—something managed through attention, process, or reaction—authority degrades under load. When safety is treated as a property of the system itself, authority persists even as conditions worsen.

This distinction separates systems that tolerate power from systems that can wield it.

Overspecification, in this light, is not an indulgence. It is an act of discipline. It simplifies design by removing debate. It stabilizes behavior by refusing to negotiate with extremes. It allows engineers to focus on what remains uncertain rather than repeatedly defending what should never have been in question.

The Fleetwood program depended on this discipline.

Before torque could dominate, control had to dominate torque. Before performance could be explored, safety had to be settled. Before ambition could expand, authority had to be fixed.

Only then could power enter the system without distorting it.

The brakes were not oversized to be impressive. They were oversized so they would never need to be discussed again.

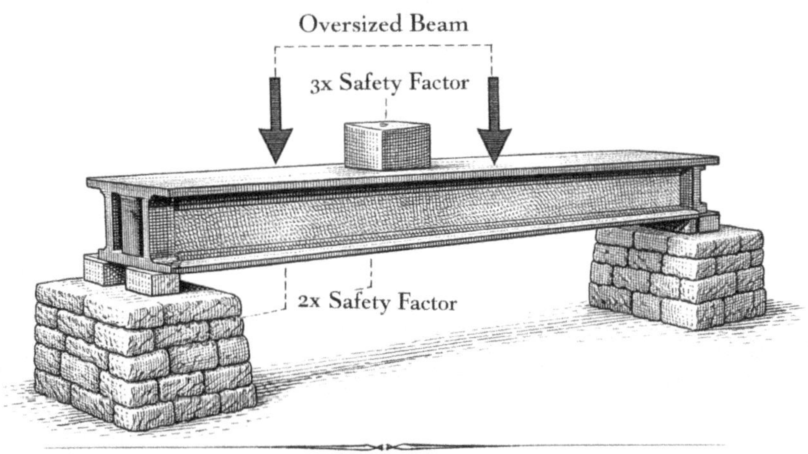

Margin is authority preserved.

12

Reducing the Unknown Surface Area

Known-Knowns as Structural Inheritance

By the time the Fleetwood program reached this phase, many of the most consequential decisions had already been made—not through enthusiasm, but through discipline.

ZF. Bilstein. Bosch. Recaro. Connolly.

These names did not surface as candidates. They surfaced as constraints.

They were not defaults. Defaults are what teams accept when no one wants to argue. These solutions were chosen precisely because argument was unnecessary. Their behavior was not hypothetical. Their operating envelopes were known, their failure modes mapped—often under conditions far harsher than anything this project would impose.

What mattered was not familiarity. It was **excess envelope.**

Each inherited solution had already lived beyond the demands of the Fleetwood. They had been loaded harder, stressed longer, and forced to behave under duty cycles well outside the domain being defined. Their limits lay comfortably beyond the system's needs. That asymmetry is where authority comes from.

A solution that merely meets requirements remains fragile. A solution whose failure boundary lies beyond relevance becomes invisible.

This is the difference between predictability and trust.

Because these behaviors were known—and known past the point of interest—debate collapsed naturally. Engineers were not speculating about how components might behave. They were working from how they had behaved, repeatedly, under conditions that dwarfed the present use case. Entire classes of questions never arose.

Inheritance, in this program, was therefore not about taste or tradition. It was a structural maneuver.

Each inherited solution arrived carrying history: operational data, revision cycles, failure analysis, and lessons already paid for elsewhere. These were not blank slates. They arrived with boundaries—and those boundaries mattered. They defined where behavior was predictable, where degradation was graceful, and where failure occurred. More importantly, they defined regions the Fleetwood would never approach.

Every inherited solution collapsed a portion of the unknown surface area.

Testing scope shrank immediately. Validation shifted from discovery to confirmation. Engineers did not need to invent new strategies or simulate hypothetical edge cases; the relevant questions had already been answered, often under harsher conditions. Emergent failure risk dropped as well. Stabilized components introduce fewer unknown interactions than systems designed together from scratch.

This is the difference between structural inheritance and cultural reuse.

Cultural reuse copies patterns because they feel familiar. Structural inheritance accepts a solution whole—including its constraints—and then designs deliberately within them. One is mimicry. The other is engineering.

The Fleetwood program practiced the latter.

Inherited solutions were not treated as negotiable. Their limits were not optimized away or reinterpreted. They were accepted as facts. Downstream decisions respected those facts rather than challenging them.

Authority was inherited intact.

This distinction matters: authority can only be inherited if its limits are accepted along with its capabilities. Rejecting those limits reintroduces uncertainty under the guise of familiarity. What was once known becomes conditional again.

When uncertainty disappears at one layer, stability propagates to the next. Suspension tuning becomes about character rather than compensation. Braking integration becomes about feel rather than capacity. Electrical systems become infrastructure rather than vigilance. Interior decisions become ergonomic rather than durability risks.

Most importantly, engineering attention is preserved.

Attention is finite. Every hour spent revisiting a solved problem is an hour not spent confronting the problem that actually matters. By inheriting authority where it already existed, the program protected attention for the one domain where truth had not yet been established.

The engine.

This pattern is routine in mature physical engineering cultures. It is far less common in software.

In software systems, components are often chosen because they function, not because their failure envelopes are irrelevant to the problem at hand. Proven libraries are replaced with bespoke frameworks. Stable infrastructure is abstracted into novelty. Stability is mistaken for stagnation, and familiarity for complacency.

The cost is rarely immediate. It emerges later as fragility, surprise, and expanding blast radius under load. Systems become busy defending themselves rather than doing their job.

The Fleetwood program treated stability differently. It was not a lack of ambition. It was a boundary. Innovation was not discouraged; it was concentrated. Every act of inheritance was a deliberate subtraction—one less variable that could distort behavior under stress.

Choosing what not to innovate was not conservatism. It was focus.

By the end of this phase, much of the car's behavior had already been determined—not by creativity, but by restraint. The known-knowns did their work quietly. They demanded no attention and required no explanation.

They existed so that the remaining unknown could be examined honestly.

And with inheritance complete, only one domain was permitted to remain unresolved.

The one that would define the car.

Only One Unknown Is Allowed

Complex systems do not fail because a single decision was wrong. They fail because too many decisions were unresolved at the same time.

This was the lesson that governed the Fleetwood program as it moved from definition into execution. It was not enough to reduce uncertainty broadly. Uncertainty had to be concentrated.

Traditional engineering disciplines operate with a simple internal rule: only one first-order unknown is allowed in a system at any given time.

This was not a slogan. It was a survival mechanism.

Unknowns do not add linearly. They multiply. Each unresolved domain introduces interactions that cannot be predicted in isolation. When several of those domains are allowed to move at once, the system stops behaving like a machine and starts behaving like a rumor—difficult to verify, impossible to trust, and expensive to correct.

The antidote is asymmetry.

Everything except the unknown must be stable enough to absorb it. Not "good enough." Not "close." Stable in the sense that behavior is already understood, boundaries are already defined, and surprises are unlikely to propagate.

This was why the earlier work mattered. Known-known solutions reduced background noise. Inheritance collapsed entire categories of doubt. Jurisdiction was declared. Margins were fixed. By the time the remaining unknown was named, the system was already quiet enough to hear it speak.

That remaining unknown was the engine.

Naming it explicitly mattered. The engine was not allowed to be an emergent question. It was not hidden among a list of open items. It was elevated, isolated, and acknowledged as the only place where truth had not yet been established.

Everything else—braking, suspension, structure, electronics, interior—was deliberately stabilized so that engine exploration would not distort the rest of the system. If something behaved unexpectedly, the cause would be obvious. There would be nowhere for ambiguity to hide.

This containment transformed risk.

The engine was no longer a gamble embedded in a moving target. It became a bounded experiment operating inside a controlled environment. Questions about power, mass, and character could be asked without simultaneously questioning the frame, the brakes, or the drivetrain integration. Learning became possible because the system around the unknown refused to move.

This pattern is familiar in physical engineering. It is far less common in software.

Software projects routinely violate this rule by allowing multiple first-order unknowns to proceed in parallel. New features are introduced while architectures are refactored and infrastructure is replaced. Data models shift while deployment pipelines change. Teams call this progress.

What it actually produces is fragility.

When behavior changes, no one knows why. Failures cannot be isolated. Rollbacks become partial. Testing expands instead of contracts. Learning slows precisely when clarity is most needed.

This is why "big rewrites" fail. Not because rewriting is inherently wrong, but because too many unknowns are allowed to coexist. Innovation becomes diffuse. Attention fragments. The system loses its ability to explain itself.

Safe innovation requires containment.

It requires choosing where uncertainty is permitted to live, and refusing to let it spread. It requires discipline to say that everything else must remain still—not forever, but long enough for truth to emerge.

The Fleetwood program made that choice deliberately.

Only one unknown was allowed. It was named. It was staffed. And it was given a system stable enough to tell the truth about it.

What remained was not caution.

It was focus.

The Engine as a Bounded Experiment

By the time the engine was discussed in earnest, it no longer represented risk in the way large unknowns usually do.

It was unresolved. It was intentionally deferred. And it was tightly scoped.

That combination mattered.

The engine was not treated as a bet placed early in the program, nor as a decision postponed out of indecision. It was treated as an experiment—one whose boundaries were drawn clearly enough that failure, if it occurred, would be informative rather than destructive.

This distinction separates inquiry from gambling.

Everything else in the system had been shaped to tolerate variation. Braking capacity exceeded any plausible output. Structural rigidity was established before torque was applied. Suspension and driveline solutions were chosen from known envelopes with excess margin. Electrical and control solutions were inherited, not invented. Interior and comfort solutions were stabilized early to prevent distraction later.

The system was quiet on purpose.

Silence is a prerequisite for learning. When too many subsystems are unsettled, unexpected behavior becomes indistinguishable from noise. When only one domain is allowed to move, truth surfaces quickly.

This was the environment created for the engine.

If power delivery felt wrong, the cause would be obvious. If weight distribution distorted behavior, it would not be masked by braking fade or chassis flex. If torque overwhelmed the driveline, the failure would occur at a boundary already understood. There would be no ambiguity about where responsibility lay.

Just as important, failure was allowed—but only failure that spoke clearly.

The engine was permitted to be wrong. The system around it was not.

This containment transformed the role of engineering judgment. Decisions were no longer defensive. Engineers did not have to speculate about compounded interactions or wonder which subsystem was lying. The system had been designed to expose error, not absorb it quietly.

This pattern is well understood in mature software organizations, even if it is not always practiced consistently.

Canary systems exist to allow new behavior to reveal itself without endangering the whole. Feature flags exist to separate deployment from activation. Blast radius reduction exists to ensure that experiments fail locally rather than globally. In each case, the principle is the same: uncertainty is acceptable only when its consequences are bounded.

What is often missed is that these mechanisms only work when the surrounding system is stable. A canary deployed into an already shifting architecture tells no truth. Feature flags layered onto unresolved design decisions conceal more than they reveal. Experiments conducted without jurisdiction simply distribute risk.

The Fleetwood program avoided this by making the engine the only experiment that mattered.

It was staffed accordingly. It was isolated architecturally. It was given a system rigid enough—physically and conceptually—to report honestly on its behavior. Learning was not delayed by cleanup. Correction did not require reinterpretation.

This was not caution. It was confidence.

By the time the engine would eventually be chosen, its role would no longer be speculative. It would be the last variable introduced into a system already prepared to accept it.

The engine was not a gamble placed in hope.

It was a question asked in a room quiet enough to hear the answer.

Why Troy Exists

The decision to establish a program presence in Troy, Michigan, on the outskirts of Detroit was not administrative.

It was technical.

Troy did not exist to accelerate approvals, enforce governance, or impose hierarchy. It was not a project office in the conventional sense. No authority flowed from it downward. No decisions were ratified there that could not have been ratified elsewhere.

Its value lay entirely in proximity.

The remaining questions surrounding the engine could not be answered by specification alone. They were not matters of dimension or tolerance that could be resolved by drawing or document. They were questions of tradeoff—why one material had been chosen over another, why a certain geometry behaved better under load, why a previous solution had failed quietly rather than catastrophically.

Those answers lived in people.

They lived in engineers who had seen platforms bend under torque. In specialists who remembered why a particular casting approach had been abandoned.

In conversations that never made it into design notes because they occurred after the meeting ended, when someone admitted that the numbers were correct but the behavior was wrong.

This is the kind of knowledge that does not survive translation.

Documentation preserves conclusions. It rarely preserves doubt. Spreadsheets capture decisions. They do not capture hesitation. Design reviews record what was agreed upon, not what was almost tried and quietly rejected.

The engine questions required access to that missing layer.

They required engineers who could say not just *what* had been done, but *why* it was done—and what had been learned when it went wrong. They required the ability to challenge assumptions in real time, to sketch alternatives, to argue without protocol, and to arrive at understanding before arriving at agreement.

These interactions do not scale asynchronously.

They require shared context. They require trust. They require enough familiarity that disagreement is interpreted as curiosity rather than resistance. They require proximity.

Troy provided that proximity.

It placed AMG engineers within walking distance of platform specialists, drivetrain engineers, materials experts, and people who understood the unwritten history of General Motors' engine programs. It collapsed the distance between question and answer. It allowed ambiguity to be explored before it hardened into decision.

Most importantly, it allowed the engine to be discussed as a system with consequences, not as an abstract specification.

Software engineering often underestimates the cost of distance in this way.

Distributed teams can share documents, tickets, and code repositories. They cannot easily share intuition. When understanding is fragmented, failures recur not because knowledge is missing, but because it is dispersed. Lessons learned in one place fail to influence decisions made elsewhere. Documentation grows thicker while insight grows thinner.

Asynchronous design processes amplify this effect. Decisions appear complete on paper, yet the reasons behind them remain opaque. When conditions change, teams repeat mistakes not out of negligence, but because the original caution never traveled with the conclusion.

The limits of this approach surface only under stress.

By the time failures demand explanation, the people who understood the tradeoffs are no longer in the room—or never were.

Troy existed to prevent that outcome.

It was an acknowledgment that some truths cannot be transmitted efficiently. They must be experienced. They must be argued. They must be absorbed through repeated, informal contact.

Some truths only surface where engineers can argue over lunch.

By choosing proximity over process at this stage, the Fleetwood program accepted a simple reality: the remaining unknown was too important to be mediated through artifacts alone.

Understanding required presence.

And presence, in this case, was the most technical decision left to make.

Brand DNA as a Constraint, Not an Aspiration

For much of its history, Cadillac did not need to explain what it was.

Its identity was enforced mechanically. Engines were not interchangeable commodities or abstract performance units. They were character-defining structures. A Cadillac engine was smooth without being fragile, powerful without being demonstrative, and quiet without being inert. Torque arrived early and stayed present. Motion felt inevitable rather than urgent.

That identity was not a styling exercise. It was the consequence of engineering decisions that constrained behavior long before the car was driven.

Over time, those constraints weakened.

Platform sharing reduced structural distinction. Drivetrain consolidation blurred mechanical character. Financial optimization encouraged reuse where separation once carried meaning. None of these decisions were irrational. Each addressed a local efficiency problem. Together, they changed the role engineering played in enforcing identity.

It would be incorrect to suggest that Cadillac abandoned engine identity entirely during this period.

The Northstar V8 was a serious and ambitious engineering effort. It was Cadillac-led, technically sophisticated, and intended to reassert distinction through precision, smoothness, and modernity.

Its architecture reflected the realities and priorities of its era: emissions compliance, packaging efficiency, refinement at sustained speed, and competitive performance density.

Just as importantly, the Northstar established a numerical truth that would outlive the program itself.

It defined a weight class.

The Northstar demonstrated that a modern Cadillac V8 could deliver durability, refinement, and output within a specific mass envelope. That envelope mattered. Weight is not an abstract concern; it determines center of gravity, load paths, suspension behavior, and how torque is reacted through the structure. An engine that exceeds that mass—regardless of output—fundamentally changes how a vehicle must be engineered to feel calm rather than strained.

The Fleetwood program accepted this constraint explicitly.

The engine would not exceed the weight of a Northstar V8.

This was not nostalgia. It was discipline. The target weight was not a tribute to past architecture, but a guardrail against violating character. Any solution that exceeded that envelope would require compensatory engineering elsewhere—and compensation is the enemy of effortlessness.

The Northstar also marked a transition in how identity was expressed.

Its overhead-cam architecture concentrated complexity and mass higher in the engine. Its character favored precision and mechanical sophistication over torque-dominant effortlessness. And as it was shared internally—most visibly with the Oldsmobile Aurora—it ceased to belong to a single behavioral identity.

What had once been an identity anchor became a portfolio asset.

This was not a failure of engineering. It was a consequence of reuse.

When an engine must serve multiple brands, platforms, and characters, it can no longer enforce a single constraint. Its role shifts from defining behavior to accommodating variation. Identity moves from being structural to being descriptive.

If AMG was going to build a Cadillac, that drift could not continue.

The program could not begin by asking what components were available. It had to begin by asking what behaviors were non-negotiable. Character could not be rediscovered at the end of the process. It had to constrain decisions from the beginning.

Engineering would not be allowed to redefine identity. Identity would constrain engineering.

This inversion mattered. When brand DNA is treated as an aspiration, it becomes flexible. Compromises accumulate. Tradeoffs drift. The system adapts until no single behavior is protected. When brand DNA is treated as a constraint, decisions simplify. Some solutions are immediately disqualified. Others become inevitable.

For the Fleetwood program, this meant that certain outcomes—regardless of convenience or precedent—were unacceptable. An engine that produced impressive numbers but communicated effort was wrong. A drivetrain that favored immediacy over calm was wrong. A solution that optimized efficiency at the expense of character was wrong.

These were not aesthetic judgments.

They were technical ones.

A Cadillac does not announce power. It carries it. A Cadillac does not chase response. It absorbs demand. A Cadillac does not feel effortless because it is light. It feels effortless because effort is managed.

Those properties cannot be layered on later. They must be preserved structurally.

Software systems encounter this same tension, often without recognizing it as technical debt.

Products begin with clear intent—simplicity, reliability, trust. Over time, platform convenience and reuse compete with that intent. Decisions are justified because they align with shared infrastructure or reduce short-term cost. Gradually, the system begins to resemble every other system built on the same foundations.

Homogenization is rarely planned. It emerges incrementally, as constraints are relaxed and exceptions accumulate. What is lost is not functionality, but character. The system still works. It simply no longer behaves intentionally.

This loss has technical consequences. Performance degrades in subtle ways. User experience becomes inconsistent. Engineering effort shifts from design to compensation.

Homogenization becomes technical debt.

The Fleetwood program resisted this by treating Cadillac's identity as invariant. It was not something to be honored symbolically. It was something to be enforced mechanically. By doing so, uncertainty was reduced further.

Entire classes of solutions were eliminated without debate.

What remained had to respect the character that defined the brand long before this project began.

The engine, still unresolved, now had fewer places to hide.

It would have to deliver torque without agitation. It would have to move mass without explanation. It would have to respect a weight envelope proven possible. It would have to feel unmistakably Cadillac.

Those requirements were not preferences.

They were constraints.

The Customer as a Forcing Function

The remaining uncertainty did not resolve itself through analysis alone.

It resolved because someone was willing to commit.

The customer already trusted AMG. He understood its discipline, its restraint, and its refusal to sell ideas before they were defensible. What he wanted was not an option package or a derivative. He wanted something that did not exist—something that could not be purchased by waiting for a product cycle to complete.

That intent mattered.

Time, money, and clarity collapse indecision in a way no design review ever will. When commitment is real, ambiguity becomes expensive. Questions that linger in hypothetical space are forced into the open. Tradeoffs stop being academic. Engineering moves from possibility to responsibility.

This was not indulgence.

It was clarity through commitment.

The customer's request did not dictate solutions. It dictated seriousness. The program was no longer an exercise in exploration; it was an obligation to produce truth. That obligation reframed the remaining uncertainty around the engine. The question was no longer whether an answer could be found, but how to pursue it without contaminating the rest of the system.

Here, precedent mattered.

General Motors had recently navigated a similar problem space with the Corvette ZR-1 program. That effort had already established a workable pattern: external engineering expertise leveraged where appropriate, specialized casting capabilities brought in where required, and final responsibility retained in-house. Lotus contributed engineering insight. Mercury Marine provided aluminum casting expertise at a scale and quality level few others could match.

Crucially, Mercury Marine would not assemble finished engines for AMG.

This distinction was not procedural. It was jurisdictional.

Casting is a manufacturing capability. Final assembly is an act of authorship. It is where tolerances are enforced, decisions are finalized, and responsibility becomes indivisible. AMG had always retained that role, and it would not change here. Whatever form the engine ultimately took, it would be built, signed, and stood behind by AMG.

What the ZR-1 precedent provided was not a template, but confidence in process. The engineering relationships were known. The casting workflows were understood. The failure modes—technical and organizational—had already been encountered and addressed. This did not eliminate uncertainty about the engine itself, but it bounded the uncertainty around how that uncertainty would be explored.

That mattered.

The Troy program office now had more than proximity. It had credibility. Engineers knew which conversations needed to happen, and with whom. Introductions were not speculative. They were continuations. The path forward was not invented; it was adapted from a process that had already carried unknowns to ground without destabilizing the system around them.

The customer's commitment made this adaptation inevitable.

With funding secured and intent unambiguous, the remaining unknown could no longer be deferred indefinitely. The engine would have to be confronted directly—but only after everything else had been stabilized enough to survive the inquiry. The program had reduced uncertainty, isolated the unknown, constrained identity, fixed margins, and established jurisdiction. What remained was not risk. It was work.

Software systems encounter this moment as well.

Internal customers tolerate ambiguity longer. External customers do not. When a system has a real user—one whose business depends on it—assumptions are exposed quickly. Performance requirements stop being aspirational. Failure modes stop being theoretical. Constraints become real because someone experiences them personally.

"Real users" surface real constraints not because they are demanding, but because they are affected.

The Fleetwood program crossed that threshold here.

The customer did not prescribe the engine. He made its resolution unavoidable. By committing resources and expectation, he transformed the remaining uncertainty from an abstract problem into a bounded obligation.

Some uncertainty only resolves when someone is willing to pay for truth.

With that commitment in place, the program was no longer preparing for a decision.

It was ready to make one.

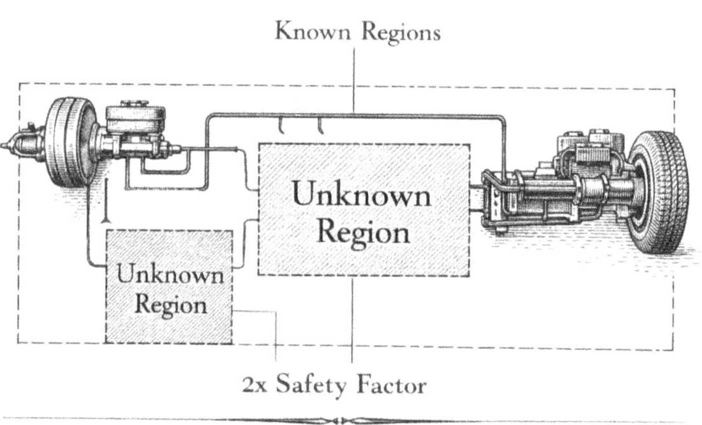

Unknowns must be isolated, not averaged.

13

Mass, Effort, and the Cost of Moving Systems

Before power could be chosen, the system had to be understood in terms of what it resisted.

Mass is often mistaken for size. Or for weight. Or for cost.

In engineering, mass is something more precise and far more dangerous: it is the effort required to change a system's state.

Mass accumulates quietly. It hides in architecture, in allocation, in assumptions made early and defended late. It is tolerated because it does not announce itself until force is applied—and by then, every correction is expensive.

Physical systems expose mass brutally. Engines mounted too high reveal themselves under load. Structures that carry weight poorly do not fail gracefully; they exhaust drivers long before they break parts. A system may meet every static requirement and still feel wrong because its mass resists change where change is demanded.

Software systems behave the same way, though the signals are easier to ignore.

State allocated too early becomes inertia. Resources reserved for possibility become load that must be carried continuously. Caches, sessions, carts, queues, and autoscaling groups accumulate mass long before they deliver value. The system still works. It simply grows heavy.

Heavy systems demand force. Force appears as hardware, as instances, as cost, as retries, as timeouts, as human intervention. Margins shrink not because they were removed, but because they were never affordable in the presence of mass.

The Fleetwood program encountered this truth before any engine was selected.

Before output could be discussed, the cost of moving mass had to be confronted—where it sat, how it shifted, and what it demanded under sustained effort. An engine that violated these truths would never feel effortless, regardless of power.

This chapter examines mass as the hidden governor of performance—physical and computational—and why reducing mass is the only reliable way to preserve margin without deception.

Power comes later.

First, the system must be light enough to accept it.

Defining Mass as Resistance to Change

Incident (Authority Definition)

An incident is any condition in which a system's declared invariants are violated or can no longer be testified to with evidence—whether or not customer harm has yet occurred.

Mass is not a moral concept.

It does not care about intent. It does not respond to justification. It does not negotiate with design narratives or delivery timelines. Mass exists whether it was chosen deliberately or accumulated accidentally, and once present, it must be carried.

In engineering, mass is often confused with weight or size because those are the forms in which it eventually becomes visible. But mass is more fundamental than either. It is not how much something weighs. It is how much effort it takes to move it, redirect it, slow it, or stop it once it is in motion.

This distinction matters because mass is rarely added all at once.

It accrues invisibly.

Small decisions accumulate. Allocations are made "just in case." State is retained because it might be useful later. Capacity is reserved because it feels safer than calculating demand. None of these choices appear dangerous in isolation. Each is locally reasonable. Together, they create inertia.

Mass does not announce itself when it is added. It reveals itself only when change is required.

A system with little mass responds immediately. Direction changes feel natural. Corrections are inexpensive. A system with high mass resists change even when change is correct. Force must be applied continuously just to maintain motion, and dramatically to alter it. When force arrives late, it is rarely precise. It compensates rather than corrects.

This is why force applied late always costs more than mass removed early.

In physical systems, this truth is unavoidable. Engineers feel it directly. A heavy rotating assembly resists acceleration and deceleration regardless of output. A high center of gravity magnifies every correction. A long polar moment turns small inputs into delayed responses. No amount of tuning can eliminate the effort required to overcome mass once it is present.

Software systems behave the same way, but with fewer sensory cues.

From the earliest stages of training, software engineers are taught—implicitly and often explicitly—that mass is cheap. Compute will get faster. Memory will get larger. Storage will become abundant. Every eighteen months, the same hardware will deliver roughly twice the capability for the same cost. What is slow today will be acceptable soon, and effortless not long after that.

This expectation is a direct inheritance of Moore's Law.

Some organizations design around it deliberately. Software is allowed to be heavy because time is expected to solve the problem. A feature that strains the system today will "age into" adequacy as hardware improves. By the time inefficiency becomes painful, a refresh cycle, platform upgrade, or rewrite is assumed to be available. Progress is deferred to physics.

What is rarely taught is the cost of carrying that mass in the meantime.

Computer science education emphasizes correctness, abstraction, expressiveness, and reuse. These are valuable skills. But engineers are not trained to reason about mass: how state accumulates, how allocations linger, how architectural decisions increase the effort required to change behavior later.

As a result, mass becomes normalized.

Code is written to be clear rather than light. Data is retained because it might be useful. Resources are allocated because they are inexpensive relative to developer time. These choices feel rational in isolation. But unlike physical systems, software systems hide the cost of mass until scale forces it into view.

In the modern era, that delay no longer exists.

Cloud computing has inverted the old assumption. Instead of buying capacity once and amortizing inefficiency over time, systems now pay continuously for every resource they consume. Memory, compute, coordination, bandwidth, and storage are no longer sunk costs. They are rent.

Mass is no longer hidden behind depreciation schedules. It appears directly on the invoice.

Heavier systems scale poorly not because they cannot grow, but because they require disproportionate force to do so. Autoscaling becomes compensation rather than correction. Costs rise faster than demand. Margins shrink. Engineers respond by adding more infrastructure rather than removing mass. The system survives, but it does not become light.

It becomes dependent on force.

Once a system requires constant force to behave acceptably, every future decision becomes constrained by that requirement. Flexibility declines. Optionality disappears. Safety margins become expensive. Performance tuning turns into triage.

This is not a failure of effort or intelligence.

Mass does not accumulate because engineers are careless. It accumulates because systems are designed to make progress before they are designed to change. Allocation precedes measurement. Possibility precedes probability. Convenience precedes discipline. Each step feels justified. The resulting inertia is not.

Physical engineering learned this lesson long ago. Colin Chapman, founder of Lotus Cars, once observed that adding power makes a car faster only on the straights, while subtracting weight makes it faster everywhere. The insight was not about speed. It was about effort. A lighter system responds more readily to every input, in every condition.

The same principle applies to software.

Reducing mass improves responsiveness under load, lowers operating cost, and makes safety margins affordable. Adding force—more hardware, more instances, more retries—can compensate temporarily, but it never changes how hard the system is to move.

This is why mass must be confronted directly, especially when it appears cheap.

Mass is any property of a system that increases the effort required to change its state.

Once that definition is accepted, everything that follows becomes mechanics.

Physical Mass in the Fleetwood

In a large car, mass is never neutral.

It does not simply sit in place. It shifts, resists, amplifies, and demands compensation. Where mass is located matters as much as how much exists. How it moves matters more than how it is measured.

The Fleetwood made this unavoidable.

This was not a small car tuned to feel large. It was a large vehicle expected to move without announcing the effort required to do so. At sustained Autobahn speeds, effortlessness is not a sensation—it is a consequence of physics managed correctly over time.

The first consideration was not output. It was balance.

Engine mass does more than add weight to a vehicle. It establishes the center of gravity around which every other system must operate. A higher center of gravity increases load transfer during acceleration, braking, and cornering. It magnifies pitch and roll. It increases the effort required for the suspension to maintain contact and for the brakes to remain predictable under load.

Architecture determines where that mass lives.

An overhead-cam engine concentrates complexity—and therefore mass—higher in the engine assembly. Valvetrain components, cam carriers, and associated structure sit above the crank centerline. The result is not merely additional weight, but elevated weight. That elevation changes the car's response even when total mass remains within acceptable bounds.

By contrast, a cam-in-block architecture places more mass lower in the system. The valvetrain is compact. The rotating assembly remains closer to the vehicle's roll center. The engine's contribution to the polar moment—the resistance to rotational change—is reduced.

These differences are not academic.

At speed, they determine how much effort is required to keep the vehicle settled. They influence how much correction the driver must apply, how much compensation the suspension must provide, and how much reserve the braking system must carry simply to remain trustworthy.

Two engines with identical output do not impose identical effort.

This is the hinge.

Horsepower figures and torque curves describe what an engine can deliver. They do not describe what the vehicle must endure to accommodate it. An engine that meets every performance target can still degrade the system if its mass is poorly located or its inertia poorly managed.

The Fleetwood's mission made this distinction critical.

Sustained Autobahn travel is not a series of transient events. It is a prolonged state. At 250 kilometers per hour, small imbalances compound. Minor corrections accumulate. Systems that feel acceptable during brief demonstrations become exhausting over time. Effortlessness disappears not because the car cannot maintain speed, but because it cannot maintain composure without continuous intervention.

This is where polar moment becomes decisive.

Mass distributed far from the vehicle's center resists directional change. The longer the moment arm, the greater the effort required to alter course, even slightly. In a large sedan, this resistance is felt not as sluggishness, but as delay. Inputs arrive late. Corrections overshoot. The driver works harder to achieve the same result.

Reducing polar moment does not make a car feel lighter in a static sense. It makes it feel calmer under dynamic conditions. The vehicle responds without anticipation or apology. Corrections remain proportional. Stability feels intrinsic rather than enforced.

The Fleetwood required this calm.

An engine that added mass high, forward, or far from the vehicle's center would demand compensatory engineering elsewhere. Stiffer suspension. More aggressive damping. Larger anti-roll measures. Each compensation would erode the very character the program sought to preserve.

Effortlessness cannot be tuned in after the fact.

It must be preserved structurally.

This is why the discussion of mass preceded any discussion of power. Output could always be increased later. Effort, once imposed, is difficult to remove. A system that carries mass poorly becomes dependent on force to remain acceptable.

The Fleetwood could not afford that dependency.

The engine—still unchosen—would have to respect these constraints. Not as guidelines. As boundaries. Whatever form it took, it would need to deliver its output without shifting the system's balance, raising its center of gravity, or increasing the effort required to maintain composure at speed.

Only engines that satisfied these conditions were even eligible for consideration.

Everything else—no matter how impressive on paper—was already wrong.

Software Mass and Premature Allocation

Software mass rarely announces itself as cost.

It announces itself as inertia.

Systems feel slower to change. Adjustments take longer than expected. Small improvements require disproportionate coordination. Performance tuning produces marginal gains that evaporate under real load. Engineers apply more force—more instances, more replicas, more buffering—not because the system cannot function, but because it cannot respond.

This resistance is not accidental.

It is the result of premature allocation.

In software, allocation often precedes behavior. Resources are reserved before demand is known. State is created before it is earned. Capacity is provisioned for what *might* happen rather than what *does* happen. Each allocation feels prudent. Each reduces immediate uncertainty. Together, they harden the system long before its true operating characteristics are understood.

This is where mass forms.

A clear example appears in e-commerce systems.

Carts are often allocated eagerly. Every visitor receives one. Session state is established immediately. Storage is reserved. Background processes begin tracking activity that may never occur. The system behaves as if every user is about to transact, even though a majority never will.

This is not waste in a moral sense.

It is inertia in a mechanical one.

The system must now carry the mass of every potential cart—memory, coordination, persistence, cleanup—regardless of whether that cart is used. Even when usage rates are low, the *possibility* of use has already been engineered into existence. The system resists change because it is already busy maintaining what might happen.

Allocation has become obligation.

The same pattern appears elsewhere.

Sessions that persist indefinitely. State that survives "just in case." Queues that exist before producers arrive. Workers that idle continuously to preserve readiness. Each decision is defensible in isolation. Each creates mass that must be carried continuously.

Crucially, allocation is not free simply because it is cheap.

The cost is not just financial.

Premature allocation increases the effort required to change the system. Removing state becomes dangerous. Reclaiming resources risks breaking assumptions baked in early. Performance optimizations are constrained by obligations that were never validated against real behavior.

Engineers respond predictably: they add force.

Autoscaling absorbs peaks. Caches mask latency. Retries smooth failures. These interventions keep the system operational, but they do not make it lighter. They increase dependence on force rather than reducing inertia.

The system works—but it does not adapt.

This distinction matters.

Engineering for possibility is not the same as engineering for behavior. A system designed around what *could* happen must carry all those possibilities at once. A system designed around observed behavior can remain light, responsive, and cheap to change.

Premature allocation commits the system to a future it has not yet earned.

In physical systems, this mistake is obvious. No engineer would install suspension sized for a fully loaded vehicle before knowing whether the vehicle would ever carry that load. Doing so would degrade handling, increase effort, and require compensation elsewhere. The vehicle would feel wrong even when empty.

Software systems make this mistake routinely because the signals are quieter.

The harm appears later, under scale, under change, under stress. By then, the mass has already hardened. Removing it requires more effort than carrying it, so it remains.

This is why premature allocation is one of the most reliable ways software accumulates mass.

Not because it wastes resources, but because it creates inertia.

And inertia, once established, demands force to overcome.

The next question is not whether this mass has a cost—it does.

The question is how long systems can afford to pay it.

Cost as a Symptom of Hidden Mass

In software systems, cost is rarely treated as a diagnostic signal.

It is treated as an outcome.

Budgets are exceeded. Invoices climb. Finance asks questions. Engineering responds with explanations about growth, success, or unavoidable complexity. Cost is framed as the price of doing business at scale.

This framing misses the mechanism.

In most modern systems, cost is not the result of scale alone. It is the symptom of mass that was never removed.

Cloud computing makes this visible in a way earlier infrastructure models did not. In on-premise environments, inefficiency could be amortized. Hardware was purchased once. Excess capacity sat quietly. Mass was hidden behind depreciation schedules and sunk costs. A heavy system looked no different from a light one until it failed.

The cloud charges differently.

Every unit of mass incurs rent. Memory allocated but idle is still paid for. Instances waiting for work still consume budget. Coordination maintained "just in case" still requires effort. State retained unnecessarily still demands replication, backup, and monitoring.

Cost grows because mass was never reduced.

The pattern is consistent. State is allocated early. Resources are provisioned for possibility. Services remain resident to preserve readiness. Each choice adds mass. When load increases, the system resists. Force is applied. Capacity grows. Costs rise faster than value.

Margins shrink not because the business model is weak, but because the system is carrying more inertia than it can afford.

This is where many organizations begin to feel trapped.

Removing mass now feels dangerous. State is entangled. Dependencies are opaque. Performance work becomes incremental and defensive. Engineers focus on managing spend rather than changing structure. Cost optimization turns into rightsizing exercises that treat symptoms rather than causes.

The system becomes efficient only in the narrow sense that it no longer collapses.

This is not optimization.

It is accommodation.

Physical engineering would never accept this framing. No engineer would argue that a vehicle requiring twice the fuel to achieve the same performance is "successful" simply because it reaches the destination. Excess effort is recognized immediately as inefficiency, not growth.

Software systems avoid this recognition because the feedback loop is indirect.

Cost arrives after design decisions are made. It is mediated through dashboards and invoices rather than physical sensation. Engineers do not feel inertia in their hands; they see it in monthly reports.

By then, the mass is already structural.

This leads to an uncomfortable but necessary question.

If an organization claims to care about performance, yet shows no sustained interest in reducing resource usage—no curiosity about where mass lives, how it accumulates, or why effort keeps increasing—then it is worth asking why. Not as an accusation, but as an examination of preparation.

In many organizations, performance engineering teams are not formed deliberately. They are assembled. Individuals are promoted into performance roles from adjacent disciplines—development, operations, quality assurance—because they are competent, reliable, and available. What they are rarely given is formal training in performance as a first-class engineering discipline, or mentorship from someone who has learned—often painfully—how mass manifests in real systems.

Without that foundation, performance work drifts.

Attention shifts to tooling. Dashboards multiply. Benchmarks are run. Load tests confirm that the system survives. But survival is mistaken for efficiency, and stability is mistaken for lightness. The harder question—why does the system require so much force to behave acceptably—is never fully confronted.

This is not negligence.

It is inheritance.

Performance teams often inherit responsibility without authority, measurement without mandate, and outcomes without control over structure. They are asked to manage symptoms while the sources of mass remain out of bounds. Reducing resource usage becomes someone else's concern. Architectural weight is treated as fixed. Cost optimization becomes an exercise in adjustment rather than understanding.

In that environment, obsession with mass would be disruptive.

It would require asking questions that reach beyond tuning and into design. It would surface tradeoffs that were never made explicit. It would challenge assumptions embedded long before the performance team arrived.

So it often does not happen.

The result is predictable: systems that function, scale, and cost more than they should—because no one was tasked, trained, or empowered to make them light.

Understanding mass is not optional expertise.

It is the prerequisite for authority.

Only understanding can make a system light.

Mass Reduction Enables Margin

Margin is often treated as an indulgence.

Something added late, if time allows. Something afforded only by large budgets or conservative cultures. In software, margin is frequently framed as inefficiency—extra capacity, redundant systems, unused headroom that could have been converted into features or cost savings.

This framing reverses cause and effect.

Margin is not expensive because it exists. Margin is expensive because mass makes it so.

When a system is heavy, every unit of safety requires disproportionate force. Extra capacity must be provisioned continuously. Redundancy multiplies cost. Failover paths become complex and fragile. Safety is purchased repeatedly rather than embedded structurally.

Lighter systems behave differently.

When mass is reduced—when state is minimized, allocation is delayed, coupling is loosened, and inertia is removed—the effort required to change system behavior drops dramatically. Responses become proportional. Corrections become precise. Small inputs produce predictable outcomes.

Under load, these systems do not panic.

They slow gracefully. They shed work deliberately. They protect their core behaviors without collapsing into retries, backlogs, or cascading failure. Performance degrades in ways that are legible rather than surprising.

This is not because they are more powerful.

It is because they are easier to move.

Reduced mass makes margin affordable. Capacity no longer needs to be held continuously "just in case." Redundancy can be selective rather than universal.

Safety buffers can be sized for reality rather than fear. The system carries reserve without being dominated by it.

In physical engineering, this relationship is well understood.

A lighter vehicle can afford larger brakes without penalty. It can tolerate softer suspension without losing control. It can carry structural margin without becoming dull. Safety does not have to fight mass; it is supported by its absence.

The same dynamic applies to software.

Systems that are light can afford to be conservative. They can carry margin without cost explosions. They can absorb spikes without architectural contortions. They can remain calm under stress because they are not already burdened by unnecessary inertia.

This calm is not aesthetic.

It is operational authority.

Importantly, mass reduction does not require fragility. Removing mass is not about removing strength. It is about removing obligation. Strength remains where it is required. Structure is preserved where it matters. What disappears is the weight of unearned possibility.

This distinction matters for the decision still ahead.

The engine has not yet been chosen.

But the system into which it will be introduced has now been defined in terms that make certain choices impossible. Any engine that adds effort where effortlessness is required is wrong. Any architecture that raises the cost of correction, compensation, or stability is wrong. Any solution that demands force to preserve composure is wrong.

Those engines may be impressive.

They may be powerful.

They may even work.

But they cannot be right.

By reducing mass first, the program preserved the ability to choose power later without distortion. Margin became something the system could afford, not something it had to purchase repeatedly. Effortlessness remained a property of structure, not tuning.

The engine remains unchosen.

But the shape of the correct answer is now clear.

And everything that does not fit it has already been eliminated.

Once authority has been violated, continued motion is no longer neutral. At that point, intervention is not disruption—it is obligation.

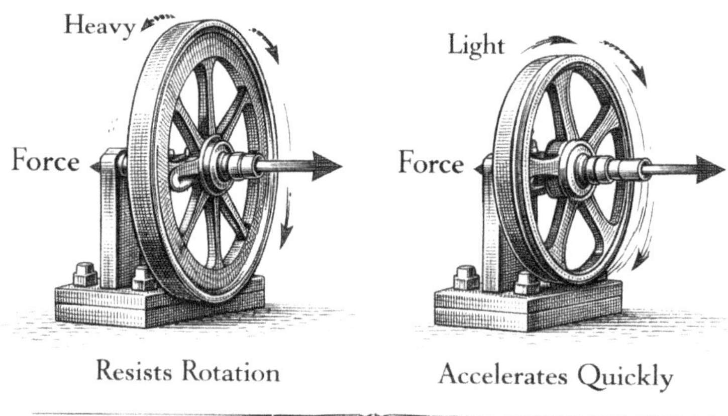

Mass governs effort, not intent.

Plans are worthless, but planning is everything.

Dwight D. Eisenhower
34th President of the United States

14

The Executive Express

Executive Warning:

Authority engineering will reduce visible velocity before it increases safety. Dashboards may worsen. Incidents may appear more frequent. This is not regression—It Is honesty returning. Executives who equate calm with correctness will resist this phase.

The last six weeks did not begin with a car.

They began with an agreement.

Cadillac pulled a Fleetwood off the line early—before the vinyl top, before the drivetrain, before any interior commitment had been made, before the VIN was applied. Basic black. Pre-wired. No indulgence. No explanation required.

The lead platform engineer from Cadillac was present, not to approve the decision, but to witness the moment the platform stopped being a finished product and became a jurisdiction.

At that moment, authorship changed hands. The platform was no longer being completed. It was being governed.

This was parity. This was how Alpina & Ruf received bodyshells from suppliers in that era. Nothing ceremonial. Nothing symbolic. Just structure released deliberately, before interpretation.

The roller left Detroit without apology.

Week One: Subtraction

The bodyshell arrived in Germany and immediately went backwards.

Interior components came out first. Dashboard. Door cards. Headliner. Anything that existed to soften, decorate, or explain was removed. This was not demolition. It was clarification. You do not reinforce what you have not measured.

Authority engineering begins by removing interpretation layers before reinforcing structure.

The interior team worked quietly. Sound insulation zones were mapped. Carpet thicknesses tested. Artificial veneers rejected without debate. Real wood would be used—American species—sourced from the same suppliers trusted by Rolls-Royce, Bentley, and Aston Martin. Proven hands. Known behavior.

Nothing new was invented. Nothing decorative was reimagined. Inherited authority was reused without apology.

Week Two: Separation

Temporary supports were installed, and the bodyshell was lifted from the chassis.

From that moment on, the car ceased to be singular. Body and chassis were sent to different engineering groups. No shared assumptions. No synchronized optimism. Each structure would be asked to confess its weaknesses independently.

No authoritative system is reinforced as a coupled whole. Independent structures must reveal their limits separately before they are allowed to agree.

The chassis group reinforced only where geometry demanded it. Not everywhere. Only where deformation would reinterpret truth. Boxing where load existed. Nothing gratuitous.

The body group worked just as selectively. Stitch welding where continuity mattered. Reinforcement only where the shell needed to contribute rather than float. Sound isolation installed deliberately—not to mask structure, but to allow the body and chassis to operate as one system instead of a collection of polite disagreements.

New body mounts were specified with a firmer formulation. More honest. Still civilized.

When the shell and chassis were reunited weeks later, they no longer negotiated. They agreed.

The old suspension remained in place for transport to paint. Nothing new would be blamed for what paint would later reveal.

Week Three: Commitment

Paint came after marriage, not before.

Once authority surfaces are fixed, public appearance becomes irreversible. Paint is commitment, not decoration.

The color was red—barely. As deep as red could go before it crossed into root beer brown. Darker than garnet. Dense. Serious. A color that absorbed light rather than reflected it.

It had presence.

The chrome remained, but disciplined. Inches reduced to millimeters. Profiles tightened. Edges sharpened just enough to respect Cadillac's language without parody. The grille remained. The crest remained. Removing either would have been like removing the Spirit of Ecstasy from a Rolls-Royce. You could do it. It would not work.

The automatic antenna remained as well. It rose and fell deliberately during quality checks. No one suggested hiding it. Some authority signals are non-negotiable.

One week before road testing, the car was unmistakable.

It looked like a Cadillac.

The drivetrain was still days away, but body, interior, and electrical teams were deep into final quality passes. The cabin was complete except for the door cards, which were deliberately held back. Too many people were still moving in and out. Too many tools. Those panels would go on last, when the car was no longer a workspace.

Cadillac's lead platform engineer arrived a few weeks before day zero to observe and understand.

He walked.

For nearly an hour.

Slowly. Silently. Taking in proportions he knew by heart, now forced to carry themselves differently. Structure he had designed, standing straighter than it ever had before.

It had grown a full undertray attached to the frame, used to both strengthen the frame and manage airflow under the car. Subtle aerodynamic items were added, such as an integrated airdam and sideskirts, so well integrated into the body design that you would not know they were not part of the original design.

Eventually, he wandered into the engine shop with one of AMG's lead engineers.

They arrived just in time.

Week Four: The Heart

The engine was already running on the test stand.

The dyno room was crowded. AMG engineers shoulder to shoulder with engineers from Bosch and Edelbrock. No spectators. No ceremony. This was work.

Initial runs confirmed the fundamentals. Oil control. Temperature behavior. Load acceptance. The engine did not surge. It did not hunt. It loaded cleanly.

It was not finished.

Over the course of the week, two additional camshaft grinds were evaluated. Each change small. No theatrics. Lift and duration adjusted to flatten delivery, not chase peaks. This engine was not being built for dyno sheets. It was being built for authority at speed, with four adults aboard and silence in the cabin.

By the end of the second week, the curve stopped moving.

One point two pounds of torque per cubic inch. Redline at fifty-five hundred. Flat from fifteen hundred onward. Pump gas. Eighty-nine octane.

Six hundred foot-pounds. Eight hundred thirteen newton-meters.

The number sobered the room.

This exceeded the Hammer by two hundred newton-meters. Power was no longer the question. Translation was.

In authority engineering, power is not tuned first. Power is introduced to reveal where structure is lying.

AMG's current torque converters were now the limiting factor.

Within days, a revised converter arrived. Reinforced. Rated for one thousand newton-meters. No one suggested detuning the engine to suit existing hardware.

The engine was not adjusted to suit the system. The system was forced to rise to meet the engine.

The engine was run again.

Nothing changed.

The builder signed the block.

That signature mattered.

Week Five: Assembly

The five-speed ZF transmission and the new torque converter were mated to the engine. A second Mercury Marine block had already served its purpose as a measurement artifact. Interfaces were proven. Nothing was assumed.

The completed powertrain was lowered into the chassis.

It fit.

No shims. No persuasion. No negotiation. The mounts accepted load without protest. The driveline rotated freely. Plumbing and wiring followed, routed for inspection as much as function.

Authoritative systems hide nothing they may later need to defend.

AMG-specific suspension and brake components replaced the originals. Bilstein dampers installed and set conservatively. Brembo calipers clamped GM discs without drama.

The wheel shop contributed AMG monoblock wheels. Tire choice favored sidewall over spectacle. Sixty-series. Z-rated.

Inside, the interior team worked quickly and quietly.

Four Recaros, reskinned in dove grey. Four executive chairs. A near-full center console spine running front to back. Veneers installed and immediately covered with protective film. Blaupunkt fitted an eight-speaker system supporting radio, CD, and an interrupt for the Motorola phone. Microphones embedded in the headliner for all four occupants.

The gear selector stood at the center of it all.

Gated. Machined. Precise. Familiar to anyone who had spent time in a Mercedes of the era.

Door cards stayed off.

Always last.

Authoritative systems seal their interfaces only after internal truth is no longer negotiable.

Day Zero: Stillness

The Cadillac engineer took the driver's seat first.

He settled in, adjusted the mirrors, and rested his hand on the center console. It stayed there longer than necessary.

The shifter gate was unmistakable. It did not belong to Cadillac's tradition of column levers and effortless anonymity. This one asked to be acknowledged.

He stepped back out of the car and looked at it from the open door.

"That," he said, nodding toward the shifter, "is going to confuse some people."

The AMG engineer smiled—just slightly.

"Only until the car moves."

The Cadillac engineer considered that for a moment, then got back in and closed the door.

Day Zero: Motion

The car rolled out on its own power.

Customer delivery was still two weeks away.

AMG had access to every controlled testing surface imaginable—slopes, broken roads, water obstacles, skidpads. Those existed to compress time, to force failure into view.

But engineers also carry their own maps.

Roads learned over years. Corners that expose geometry. Braking zones that tell the truth. Lay-bys where engines are shut off and people stand quietly, listening to metal tick and cool.

On that first day, Cadillac and AMG traded the driver's seat repeatedly. Five times. Six. No ceremony.

The engine was different from any current Cadillac engine, yet instantly recognizable. Torque arrived early and stayed. It did not surge. It moved.

It felt less like a sports sedan and more like a locomotive.

The numbers disagreed.

By any external measure, the car sat squarely in the middle of AMG's current offerings. Sports-car territory everywhere else in the world. But from behind the wheel, it did not feel athletic.

It felt inevitable.

The cabin remained quiet through most events. The radio stayed off during the Autobahn run. No one wanted to bend the antenna.

At one hundred sixty-five miles per hour, aerodynamics spoke clearly.

No more.

Authority is not the ability to exceed limits. It is the ability to recognize them without protest.

The car accepted the boundary.

The tires still had more to give. So did the brakes.

That asymmetry was intentional.

Week Six: Refinement

For the next two weeks, the internal test drivers took over.

No heroics.

Suspension adjustments were made in small increments. Damping refined to remove interpretation. ABS aggressiveness softened slightly—not to reduce safety, but to improve honesty at the pedal. Shift programming tuned to respond to intent rather than urgency.

Nothing was rewritten.

Governance is not change velocity. Governance is removal of ambiguity.

Everything was clarified.

By the end of that period, the car had become something neither Cadillac nor AMG could have fully envisioned at the outset.

Not a flagship. Not a compromise. Not a statement.

An executive express.

A machine that did not ask permission. A system that did not explain itself. A vehicle that moved through space with authority rather than speed.

It was not exciting.

It was correct.

Closing

This is what Authority Engineering looks like when it is no longer theoretical.

Structure is fixed first. Power is introduced second. Interfaces are sealed last. Inherited authority is reused without apology. New authority is engineered only where it cannot be borrowed.

Nothing moves unless it must. Nothing is softened to preserve appearances.

The Fleetwood left Germany as an AMG.

It carried an AMG VIN. It carried Cadillac scale. It carried German discipline. It carried American confidence without ornament.

Nothing about it required defense.

It simply existed.

And in doing so, it rendered the argument unnecessary.

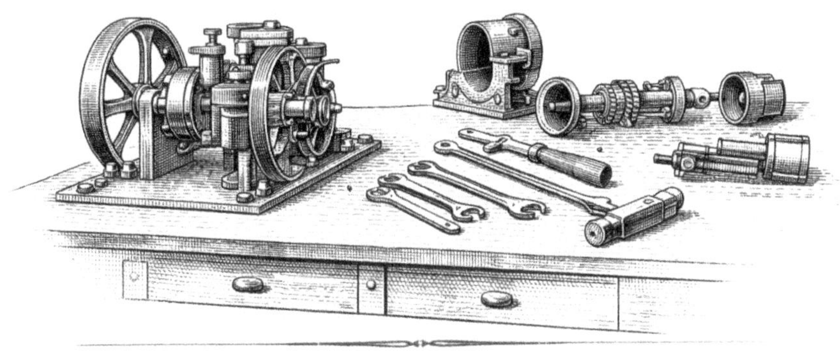

Execution is earned through preparation.

The Car That Never Existed

For thirty years, the car did not exist.

There were no photographs. No press releases. No registry entries. No footnotes in corporate histories.

Those who had worked on it did not deny it. They simply never mentioned it. Not out of secrecy—out of accuracy. The car had not been built to be seen. It had been built to be *correct*.

In the summer of 2022, that changed.

The invitation did not say much. A private placement. A single vehicle. Shown without explanation. No signage. No placards. No timeline. Just a Fleetwood, deep red, nearly brown, parked in a glass pavilion at the Pebble Beach Concours d'Elegance as part of a collection of pre-Mercedes-ownership era AMG cars, quietly asserting mass among machines that had spent decades cultivating myth.

People walked past it twice before stopping.

It did not announce itself as rare. It did not perform nostalgia. It did not ask to be understood.

Those who knew, knew.

The Reaction

A retired engineer from AMG stood nearby for most of the afternoon. He did not introduce himself. When asked whether the car was authentic, he answered simply:

"We didn't build it to be talked about."

Later, when pressed about why it had been built at all, he paused.

"Because sometimes you need to see whether a system can be right without being visible."

He smiled slightly. "It turns out it can."

Mercury Marine

For Mercury Marine, the car represented a moment most of the company never knew occurred.

One of the engineers who had overseen the aluminum block casting—long since retired—was flown out quietly. He walked around the car once. Then again.

"At the time, we thought we were just solving a materials problem," he said. "Lightweight. Strength. Thermal stability. Another application."

He rested his hand briefly on the fender.

"But that project changed how we thought about responsibility. When an engine carries that much torque, you stop thinking in terms of specs. You start thinking in terms of consequences."

He laughed softly.

"After that, we never looked at 'overbuilding' the same way again."

General Motors

At General Motors, the car had lived as an internal anomaly—never documented, never denied.

The Cadillac platform engineer who had been there at the beginning was invited as well.

He did not approach the car immediately.

"We design platforms to survive compromise," he said. "That project forced me to ask what would happen if compromise simply... didn't arrive."

He gestured toward the car.

"I watched something I helped design get pulled out of its intended future and subjected to a much harsher definition of truth. It didn't break."

He paused.

"That experience changed how I approached every platform review after that. Less tolerance for ambiguity. More respect for structure." Then, almost as an aside: "It's still a Cadillac."

AMG

For AMG, the car had never been a product.

It had been a question.

One of the engineers involved in final assembly, now long removed from day-to-day work, summed it up simply:

"That car taught us restraint."

He looked out across the field.

"We were already known for power. That project reminded us that authority isn't volume. It's refusal."

He nodded toward the Fleetwood.

"We learned that you don't always need to prove something publicly for it to be real."

The Quiet Shift

No one at Pebble Beach applauded when the car was rolled back onto its transporter.

There was no announcement that the showing had concluded. People simply noticed that it was gone.

Later that evening, a younger engineer, too young to have been involved, was overheard explaining the car to someone else:

"It's not famous because it didn't need to be."

That was the most accurate description offered all day.

Afterward

In the years since, fragments of the project have surfaced, not as stories, but as tendencies.

A little more conservatism in torque translation. A little more insistence on structural honesty. A little less appetite for explaining decisions after the fact.

The car did not create a lineage.

It created a standard.

And standards, when they are real, do not need monuments.

They simply wait— quietly— until someone else realizes they are necessary.

Without data, you're just another person with an opinion.

W. Edwards Deming
Industrial Engineer, Statistician

Appendix: Authority Engineering

What This Book Actually Argues

This book is not about cars.

It is not about software.

And it is not about tools, practices, or methodologies—though all of those appear along the way.

This book is about authority.

Not authority as hierarchy or control, but authority as responsibility for how systems behave *over time*, *under identity*, and *under stress*—when they are no longer polite, no longer fresh, and no longer operating inside ideal conditions.

Across physical engineering and software engineering, the same truths appear again and again. They do not announce themselves as philosophy. They emerge as consequences.

This appendix names those truths plainly.

Authority Precedes Optimization

No amount of tuning compensates for unclear ownership.

Systems behave according to the physics they are allowed to obey. When no one is explicitly responsible for those physics—throughput, latency, stability, cost, failure behavior—optimization becomes cosmetic.

In physical engineering, this is obvious. A chassis that flexes cannot be tuned into honesty. In software, the same mistake is easier to make. Teams optimize code paths, scale infrastructure, and add tooling while avoiding the harder question:

Who owns how this system behaves when conditions change?

239

Authority engineering begins by answering that question before any optimization begins—and by keeping that ownership intact as the system evolves.

Time Is a First-Class Engineering Dimension

Time is not a background variable.

It is an active force.

Systems that behave acceptably in the moment often fail across duration. Heat builds. Drift accumulates. Assumptions age. Latency compounds. Costs surface. Trust erodes. A system that performs well briefly but degrades quietly over time is not performant—it is deceptive.

Physical engineering treats time honestly. Fatigue, wear, and sustained load are modeled explicitly. Software systems often do not. They are validated in snapshots, benchmarks, and tests that freeze time rather than expose it.

Authority engineering insists that systems be judged not just by what they do, but by *how they behave across time*—under sustained load, repeated failure, and prolonged operation.

If time is not part of the design conversation, failure is merely delayed.

Identity Is a Technical Constraint

Identity is not branding.

It is behavior made predictable.

In physical systems, identity constrains design. A Cadillac does not behave like a sports sedan because it should not. A Rolls-Royce does not announce effort because effortlessness is part of its contract. Identity defines acceptable tradeoffs before engineering begins.

Software systems often lose identity through reuse, abstraction, and platform convenience. Shared components, shared architectures, and shared incentives dilute behavior until systems become interchangeable and incoherent.

Authority engineering treats identity as a constraint, not an aspiration.

A system that cannot explain *what it is* will eventually fail to explain *why it behaves the way it does.*

Margins Are Designed, Not Added

Safety is not something you purchase late.

It is not headroom bolted on after the fact, nor redundancy layered on top of fragility. Safety exists only where it is structurally affordable.

In physical systems, margins are baked into geometry, materials, and load paths. In software systems, margins must be baked into architecture, allocation discipline, and behavioral limits.

When margins are treated as optional, they become expensive. When they become expensive, they are cut. When they are cut, systems become brittle.

Authority engineering treats margin as a first-order design concern—not a luxury, not an afterthought, and not a line item to be negotiated away.

Mass Governs Behavior

Mass is not weight.

It is resistance to change.

In physical systems, mass reveals itself immediately. In software systems, it accumulates quietly—through state, allocation, coupling, and obligation. It remains invisible until change is required, at which point force replaces control.

Heavy systems do not fail more often. They fail more expensively. They require continuous compensation to behave acceptably. They convert effort into cost, latency, and fragility.

Authority engineering insists on reducing mass early, not because efficiency is virtuous, but because *lighter systems can afford margin*, adapt under stress, and remain legible when conditions change.

Unknowns Must Be Isolated

Innovation is not free.

Unknowns compound risk multiplicatively when they are allowed to spread. Systems that attempt to innovate everywhere at once do not become advanced; they become noisy.

In disciplined engineering, unknowns are isolated deliberately. They are bounded, staffed, and given clear jurisdiction. Everything else is stabilized to absorb their variance.

This is as true for engines as it is for software platforms, architectural rewrites, or large-scale feature introductions.

Authority engineering does not eliminate uncertainty.

It contains it.

Cost Is a Signal, Not a Surprise

Cost does not appear randomly.

It emerges from effort applied continuously to compensate for mass that was never removed. In modern software systems, cloud cost is not merely operational expense—it is feedback.

Autoscaling, orchestration, and elasticity apply force. They keep systems alive. They do not make systems light. When cost rises faster than value, it is rarely a sign of success alone. It is often a sign that the system is paying rent on ignorance.

Authority engineering treats cost as a diagnostic signal—one that points directly to where mass lives and understanding is absent.

Tools Do Not Grant Authority

Tools amplify intent.

They do not replace it.

Pipelines do not confer ownership. Dashboards do not confer understanding. Elastic infrastructure does not confer discipline. These tools are powerful, but they obey the authority already present in the system.

Where authority is unclear, tools magnify confusion. Where authority is established, tools become leverage.

Authority engineering begins before tools are chosen—and remains valid after tools change.

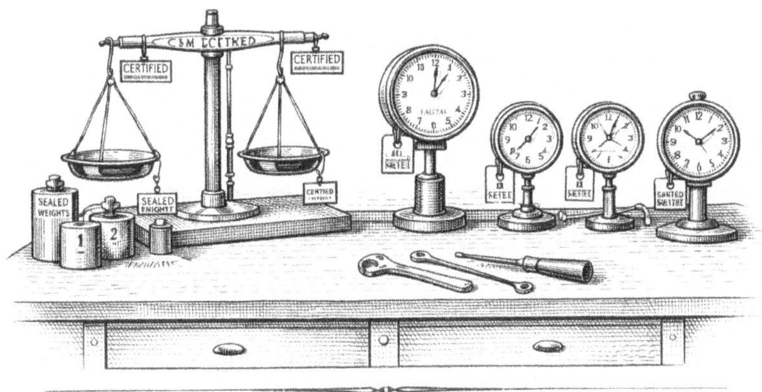

Authority precedes optimization.

Intellectual Lineage / Inspired By

This book does not present itself as a work of scholarship in the conventional academic sense. It advances its arguments through engineering analogy, structural reasoning, and observed system behavior rather than citation or literature review.

That said, its posture aligns with a long tradition of thought that treats **structure, constraint, and refusal** as prerequisites for truth. Readers familiar with the following works may recognize shared concerns and resonant themes, though no claim of derivation or dependence is made.

Engineering, Failure, and Structural Truth

- *To Engineer Is Human* — Henry Petroski

- *The Evolution of Useful Things* — Henry Petroski

Measurement, Paradigms, and the Limits of Consensus

- *The Structure of Scientific Revolutions* — Thomas S. Kuhn

Design, Affordance, and Quiet Failure

- *The Design of Everyday Things* — Don Norman

Reliability, Authority, and Organizational Failure

- *Managing the Unexpected* — Karl E. Weick and Kathleen M. Sutcliffe

Legibility, Governance, and Institutional Illusion

- *Seeing Like a State* — James C. Scott

Machines as Epistemological Instruments

- *Zen and the Art of Motorcycle Maintenance* — Robert M. Pirsig

Design Aesthetic

- *The Creative Act: A Way of Being* — Rick Rubin

The unifying thread across these works is not domain or discipline, but orientation: the belief that **systems tell the truth only when their structure refuses to negotiate under load.**

That belief is the foundation on which this book stands.

Acknowledgments

The author acknowledges the time and critical attention of colleagues and practitioners who reviewed portions of this manuscript during its development. Their feedback helped identify areas of ambiguity, overstatement, and omission.

Responsibility for the arguments, definitions, and conclusions presented here remains solely with the author.

No endorsement or agreement should be inferred.

Mike Hawkins
Rachel Laney
Mike Staten
Jeff Stewart

INDEX

About the Author

James L. Pulley III was born in Spartanburg County, South Carolina. By the sixth grade, he had lived in four cities and learned something that would shape everything that followed: that people can live next to the ocean and never see the beach. Proximity is not awareness. That observation became a career.

He graduated from Furman University in 1991 with a blended degree in computer science and business administration, and began his career at Microsoft, answering questions on the phone for Product Support Services. He became one of Microsoft's early electronic services voices on CompuServe. As an electronic services pioneer, this was the first of many times he would find himself explaining complex systems to people who needed help more than theory.

On April 1, 1996, he was hired as a field sales engineer at Mercury Interactive, the company that defined the performance testing industry. That date, and the discipline it introduced him to, became the origin coordinate of a thirty-year career in software performance engineering. April 1st was chosen for the launch of this book as the thirtieth anniversary of this hire date.

Over three decades, his work has spanned performance testing, capacity planning, application performance management, observability, and site reliability engineering. He has tested systems that governed a trillion dollars in mortgage assets, identified architectural flaws worth millions in monthly revenue, helped secure biometric entry records for every non-citizen entering the United States, and scaled a national digital wallet from ten thousand to one million concurrent users in four weeks. He has built teams, designed governance frameworks, and served as the person organizations called when the system was failing and the stakes were measured in minutes.

But the work he considers most important has always happened in the margins.

In high school, a neighbor's son lost a leg in a tragic accident during his junior year. James was asked to serve as his geometry tutor during recovery. That young man now builds custom prosthetic limbs for others who face the same challenge. The helped became the helper. It was the first time James saw the cycle that would define his professional life, but not the last.

On September 12, 2001, he began answering questions from peers in online technical forums. Not as strategy. Because it was the only useful thing available. Those answers, and the thousands that followed across the next two and a half decades, exceed the length of Homer's Odyssey. They are permanent. They are public. And they are the foundation on which every book, every framework, and every reputation in this body of work was built.

In 2008, he founded Journeyman Publishing — named for the guild tradition in which a journeyman is one who has completed an apprenticeship and earned the right to travel independently, practicing the craft wherever it is needed. He chose the title deliberately. While some of his peers see him as a master, he sees himself as a permanent learner on a journey that does not end.

In 2012, he co-founded PerfBytes, a podcast on software performance engineering inspired by the humor of CarTalk, the accessibility of Alton Brown's Good Eats, and the timing of British comedy. The show has been downloaded more than 177,000 times and continues today. It is proof that technical depth and human warmth are not opposites.

He serves on the board of the Advancing Technology Ventures Lab, a 501(c)(3) founded by Dennis Hayes to mentor early-stage ventures in upstate South Carolina. He has led, mentored, taught, and answered questions in every role he has held. In many cases, this was a task he was never asked to fill.

He often jokes that if he could extract a few cents from every success he has helped others achieve, he would never have to work again. When his father passed in 2006, he received handwritten letters from people around the world whose careers he had quietly shaped. Those letters confirmed what the work had always told him: that usefulness, practiced consistently, is the only authority that survives.

He is the author of *Navigating Performance Engineering*, *Software Performance Risk Management*, *The Fleetwood*, and the *Reputation Engineering* education series. His earlier work includes *Interviewing & Hiring Software Performance Test Professionals*, as well as dozens of articles and whitepapers across forums and conferences.

He lives in Spartanburg County, South Carolina, with his wife Rachel, in a home built around a log cabin from the late 1700s that they are expanding together, by hand. Rachel has been on this land, purchased by her parents, since she was six years old. At the same time James wandered sixteen cities across the country between 1976 and 2018 before arriving at the same three acres. Dorothy had it right in the *Wizard of Oz*, "There's no place like home..."

James has never believed that expertise is the point. The point is the person standing next to you who doesn't have it yet.